Advanced Transport and Spatial Systems Models

Tschangho John Kim
In Association with Sunduck Suh

Advanced Transport and Spatial Systems Models
Applications to Korea

With 27 Figures

Springer-Verlag
New York Berlin Heidelberg
London Paris Tokyo Hong Kong

Tschangho John Kim
University of Illinois
Department of Urban and Regional Planning
Champaign, Illinois 61801

Library of Congress Cataloging-in-Publication Data
Kim, Tschangho John.
Advanced transport and spatial systems models : applications to Korea / Tschangho John Kim.
p. cm.
Includes bibliographical references.
1. Transportation and state—Korea (South)—Mathematical models.
2. Transportation—Korea (South)—Planning—Mathematical models.
3. Mixed economy—Korea (South) I. Title.
HE277.5.K46 1990
388'.068—dc20 90-9654

Printed on acid-free paper.

Camera-ready copy provided by the author.
Printed and bound by Edwards Brothers, Inc., Ann Arbor, Michigan.
Printed in the United States of America.

9 8 7 6 5 4 3 2 1

ISBN 0-387-97277-3 Springer-Verlag New York Berlin Heidelberg
ISBN 3-540-97277-3 Springer-Verlag Berlin Heidelberg New York

To Dr. Choong Hoon Park,
who devoted his life to the modernization of Korea

Preface

The usual view in a mixed economy is that some goods and services are produced privately and some, such as transportation, are produced publicly. Private institutions, such as households and entrepreneurs, produce and consume goods and services in pursuing their parochial interests, while the public sector attempts to broaden public interests. More precisely, the public sector constructs new transportation systems, improves their capacities, and regulates services and prices; and the private sector chooses locations of production, modes of transportation, and routes of shipment. At the same time, all forms of transportation influence our lives and cause us concern for our environment, health, and safety. Thus, transportation is intimately woven into the daily life of individuals and organizations in our society. Because of its constant presence, transportation is easy to overlook until it fails in some way.

Few would contend that private firms could or should construct an efficient transportation system in a mixed economic system. Because the entire transportation system must be integrated and coordinated, firms with the power to construct such a system would have considerable monopoly control. Thus, public sector intervention is required to construct and maintain an efficient urban, regional, and national transportation system. An important question arises here. How much public intervention should there be?

The question becomes more critical in developing countries, where the existing transportation system is often inadequate and where socio-economic development goals are often established but are difficult to achieve because of limited resources. Thus, the critical issue is not only the appropriate amount of public intervention but also the type and location of services that government should provide.

A wide range of books on transportation system models is available today for the student of urban planning, geography, regional science, operations research, and economics. Yet, there are few, if any, books that deal with the question of how much a government should invest in providing transportation systems that enhance the market system. Moreover, there are at most only a few books that discuss transportation integrated with economic development strategies, particularly from the operational viewpoint.

The main thesis of this book is that the characteristics of socio-economic activity that best describe mixed economic systems are (1) the interaction between the public and private sectors and (2) the derived nature of transportation demand. On this basis, models that are introduced in this book explicitly describe real world situations where public and private sectors play significant roles in shaping regional economic systems. Further, these models are capable of suggesting optimal levels of government investment in transportation that most enhance the market system. Thus, models in this book are advanced in both theoretical and operational senses.

Theoretical issues in modeling integrated approaches are examined at length. This book's focus, however, is on modeling procedures and applications of integrated models to the analysis of real world problems. The models are actually applied to analyses of policy issues in Korea, where a strong mixed economic system prevails. It is hoped that the book will be of value to a variety of readers: urban planners, regional scientists, geographers, economists, systems analysts, operational researchers, applied mathematicians, and social scientists who are concerned with models of, and theories about, transportation in mixed economic systems.

The book is organized into five parts. In Part I, the introduction describes overall development plans and planning issues in Korea. Chapter 1 deals with two broad subjects: an overview of national development plans and issues in planning for development. Chapter 2 provides an overview of the transportation system in Korea. Part II introduces the integrated regional systems model formulated in a linear programming framework. Modeling procedures, as well as detailed assumptions involved, are explained in Chapter 3. Issues in implementing linear programming models and lessons learned from a Korean application are described in Chapter 4. In Part III, which introduces nonlinear programming models and applications, Chapter

5 describes procedures for building a nonlinear integrated regional transportation systems model in detail, and Chapter 6 describes model applications at length, particularly the formulation of optimal highway investment strategies. Part IV develops bilevel programming models of integrated transportation systems and describes their applications. Chapter 7 introduces bilevel programming approaches. Algorithms for solving bilevel programming models are described in Chapter 8. Application of the model to the estimation of highway congestion functions in Korea is shown in Chapter 9, and application to the evaluation of alternative highway investment strategies is described in Chapter 10. Part V concerns the future direction of advanced transport planning models. An alternative approach for modeling explicit public-private interaction in a mixed economic system is introduced in Chapter 11. Chapter 12 describes a research agenda in transportation in the future.

This book is a collection of works that I have researched in the years since 1978, when I directed a fine group of transportation researchers at the Regional Development Research Institute, the Korea Institute of Science and Technology (KIST). Our research team was charged with recommending a set of transportation plans that would meet the national transportation demand in the year 1991.

The bulk of the research was funded by the World Bank, the Korea Ministry of Transportation and the National Science Foundation (SES-8318239, 85-KS-05, INT-8701780, SES-8712173, SES-8718146), but it was also partially supported by the Korea Science and Engineering Foundation, the Research Board of the University of Illinois at Urbana-Champaign, the Korea Traders' Scholarship Foundation, the International Cultural Society of Korea, the Korea Maritime Institute and the Korea Research Institute for Human Settlement. I am very grateful for their support.

Permission to quote and to reproduce figures, tables, and materials has been kindly granted by Springer-Verlag, the Ohio State University Press, Kluwer Academic Publishers, and Pergamon Press. I must thank editors and publishers of the several journals involved, as well as my co-authors of articles, David E. Boyce, Sang Chuel Choe, Geoffrey J.D. Hewings, Jong Gie Kim, Chang Ho Park, and Jeong Hyun Rho, who graciously permitted me to reprint portions of articles that I co-authored.

I am indebted to my former teachers for past guidance and stimulation, especially to Edwin S. Mills, for continuing encouragement and intellectual support. I am very grateful to Jay Getz for editing and supplying detailed, invaluable comments; to Lani Leyson and Priscilla Jungmin Park for entering the text into LaTeX format; to Sang-Yun Han and Yoon Ho Yim for producing artwork and figures; and to Keechoo Choi for painstakingly formulating tables and equations according to LaTeX format during numerous sleepless nights.

Without the encouragement and understanding of my two sons, I would not have been able to finish the book by now. For reasons that I cannot describe enough, I am most of all indebted to my wife.

I dedicate this book to Dr. Choong Hoon Park, my father-in-law, who has spent most of his adult life working toward the modernization of Korea through his career as the Minister of Commerce and Industry, Deputy Prime Minister and Minister of the Economic Planning Board, Prime Minister, and Acting President of the Republic of Korea. I am deeply indebted to him for the insight and wisdom I have learned from him about planning and building a nation toward modernization, as well as for building my personal career.

Urbana, Illinois USA
March 1990

Contents

List of Tables

List of Figures

Part I

INTRODUCTION

Chapter 1

National Development Plans and Planning Issues in Korea

1.1 Introduction

The process of development in many countries is assumed to evolve through a series of stages.[1] The nature of this process depends on a number of factors, including the structure of each country's economy in terms of its endowed resources, the state of technology, the availability of entrepreneurial talent, and the individual country's position in the world economy from the perspective of the competitiveness of its products on the world market.

The Korean economy is a particularly interesting one to evaluate because it has moved far more rapidly to a stage of intermediate development than one might have anticipated in 1960s. One of the important issues stemming from the rapid economic growth is the problem related to the unbalanced distribution of socio-economic activities over space. More specifically, it centers around problems related to the centralization and decentralization of socio-economic activities.

The most commonly accepted theory regarding distribution suggests that without government intervention it is unlikely that growth centers' attractiveness to the private sector will be diluted significantly over time. If decentralization is to take place, whether for equity or even national efficiency

[1] I am indebted to Professor Sang Chuel Choe and and Dr. Sung Woong Hong for their valiable comments on the earlier draft of the Chapter.

reasons, some form of intervention by the central government will have to occur. The key question in Korea and other mixed economic systems, however, is to what extent the public sector should intervene.

The Korean government has attempted, over a number of years, to provide incentives for industrial relocation outside the Seoul area. Yet the results have been mixed; growth has taken place in these other areas, while Seoul continues to grow at an alarming rate. The reasons for this have yet to be fully teased from the data, but some preliminary analyses (Chun and Lee 1985[64]) offer suggestions.

1. New firms have migrated outside Seoul but still have substantial investments in Seoul.

2. New jobs created in firms outside Seoul have often gone to workers who commute to work from the Seoul region for the work week and return to Seoul for the weekend. Thus, the impact of the expenditure of wages and salaries is predominantly felt in the Seoul region rather than, for example, in Pusan, where the income is earned.

3. Government policies have focused mainly on the manufacturing primary sector to the point of excluding the service tertiary sector; but the latter is now the dominant employer in Korea (Table 1.1). The comparative advantage established by these service activities will make it even more difficult for firms located outside of Seoul to compete effectively.

Thus, the trends in industrial development and location in Korea reflect more the trends of a well-established, rather than developing, country. On the other hand, the policies initiated by the central government have not taken into account the nature of *international* and *interregional* competition and the relationship between them. Some of these plans and policies will be reviewed in the next section.

1.2 Five-Year National Development Plans

Korea has planned and completed five Five-Year Development Plans (FYDPs) between 1962 and 1986. During the first FYDP period (1962–1966), seven special planning districts were established that together cover a major part of Korea. Specific goals have been set for each district's development plan. The plans initiated in the first FYDP were completed in subsequent

	54	64	74	84
Total Population (million)	21.5	28.0	34.7	40.0
Urban Population (%)	24.5	31.2	46.9	62.4
Population in Seoul (%)	4.7	13.0	17.8	23.3
Export (Current $ in million)	21.7	119.1	4,460.4	24,445.1
Import (Current $ in million)	244.0	404.4	6,851.8	26,192.2
GNP/Capita (Current $)	70.0	103.0	535.0	1,880.0
GNP by Sector (%)				
Primary	39.4	46.8	24.5	13.7
Secondary	12.6	17.3	26.4	28.9
Tertiary	48.0	35.9	49.1	57.4
Employment by Sector (%)				
Primary	78.7	61.9	48.2	29.7
Secondary	1.6	8.8	17.8	23.3
Tertiary	19.7	29.3	34.0	47.0
Illiteracy (%)	23.2	14.7	11.6	4.0
Number of Farmers (1,000)	13,170	15,553	13,459	9,475
Number if Farmers (&)	61.1	55.6	41.1	31.0
Farm Land (ha)	87.3	88.6	94.0	108.1
Rice Produced (1,000 t)	2,559	4,257	4,858	5,816
Water Supply Coverage (%)	16.0	18.3	40.6	61.0
Water Supply/person-day (liter)	71	104	182	276
Telephone per 100 people	0.12	0.7	2.6	12.3
TV (1,000)	-	32	1,619	7,223
Refrigerator (1,000)	-	-	174	5,685

Source : Korea Times (Hangook-Ilbo), June 12, 1984

Table 1.1: Changes in Socio-Economic Indicators in Korea: 1954-1984

plans. For example, during the second FYDP period (1967–1971), the the 29.5-km Seoul-Incheon highway was opened (December 1968), followed by the opening of the 428-km Seoul-Pusan highway (July 1970). During the third FYDP period (1972–1976), the major thrust of the national plan was the construction of heavy and chemical industrial complexes. Objectives set to be achieved during the fourth FYDP (1977–1981) included the development of human and social resources, high technology, and an integrated infrastructure system. Subways have either been planned and/or partially constructed in Seoul and Pusan during this period.

Investment plans for infrastructure development during the third and fourth FYDP periods were actually based on the first National Comprehensive Physical Development Plan (NCPDP), which was promulgated in 1971. The plan contains programs that were to be implemented during the 1972-1981 period plans, which called for the allocation of about twenty-five percent of GNP for ten years.

The Comprehensive Water Resources Development Plan (CWRDP) of 1965 is another governmental effort that has affected distribution of the nation's socio-economic activities. The annual rainfalls in Korea average about 1,200 millimeters, forty percent higher than the global average. However, about seventy percent of those rainfalls, because they take place in the four months between June and September, cause floods in those months; in other months there are droughts. One of the major projects implemented under the plan is the construction of multipurpose dams in an attempt to reduce the variability in water availability.

The fifth FYDP of 1981 was revised in 1983. A transportation model that was developed for generating alternative programs in the transportation sector of the original fifth FYDP can be found in Kim (1980)[139]; and in Chapters 3 and 4 in this book. The models and planning procedures, as well as policy implications, are introduced in these chapters.

The model by Kim, Boyce, and Hewings (1983)[135] was built on a nonlinear programming framework in which the problem of crosshauling was addressed by entropy constraints. The model is reintroduced in Chapter 5. From the model solutions, recommendations are made for additional networks to transport increased amounts of commodities and passengers to meet the national target for exports.

Factors behind the revision of the original fifth FYDP were discussed by Suh (1983)[240], and implied effects of changes in the socio-economic climates on modeling procedures can also be found in Kim and Kim (1985)[140] and Chapter 4. The major implications of the revised fifth FYDP on the distribution of the socio-economic activities can be surmised from its goal of achieving balanced national development. By articulating this goal, the government has shifted from a growth pole development policy to a balanced regional growth policy. In addition to infrastructure and rural development, programs geared to the goal of balanced development included education, medical care, and telecommunication. In this plan, attention was focused on a broader range of attributes from concerns with social welfare (increasing the number of physicians) to general infrastructure (increasing the supply of housing and paved roads).

Results of implementing a series of government plans are illustrated in Table 1.1, where socio-economic indicators are compared during the 1954–1984 period. Some striking features in the table include the rapid increase of the urban population, a more than 1,100 percent increase in export amounts, constant decrease in the employment and Gross National Product (GNP) contribution by the primary sector, and the reversed trend in the tertiary sector. There have been subtle population shifts from rural to urban areas during the past thirty years. In 1984, about sixty-two percent of the total population of forty million lived in urbanized areas (cities with more than 50,000 people). In particular, more than sixty-five percent of the total population is concentrated in the capital and Yeongnam regions, representing an extremely unbalanced population distribution over space (Figure 1.1).

Economic activities are also unevenly distributed among regions. About eighty percent of the manufacturing firms in Korea are concentrated in two regions, the capital and Yeongnam. The level of concentration of socio-economic activities in Seoul can be further highlighted with reference to Table 1.2. Of all issues, uneven distribution of these activities clearly has created the most concern. The next section provides a review of some of the major government strategies to promote decentralization.

1.3 National Development Planning Issues

1.3.1 An Overview

The government's efforts to decentralize socio-economic activities from Seoul have been major central policies in the past twenty years. There have been sixteen executive orders issued by the Government of Korea concerning decentralization of activities from Seoul (Kim 1982[132], 846-849). Five cabinet-level ministries—the city of Seoul, the Economic Planning Board (EPB), the Ministry without Portfolio, the Ministry of Construction (MOC),

Socio-Economic Activities in Seoul, 1984	Percentage of National Totals
Population	23.3%
Total Employment	45.6%
Employment in Manufacturing Industry	21.0%
Employment in Civil Service	40.0%
Employment in Export Industry	55.8%
Employment in Trading Companies	90.0%
GNP Contribution	30.0%
Savings Contribution	64.0%
Banking Industries	37.0%
Bank Loan	61.0%
Medical Facilities	40.0%
Universities and Colleges	45.0%

Table 1.2: Measures of Concentration of Activities in the Seoul Region

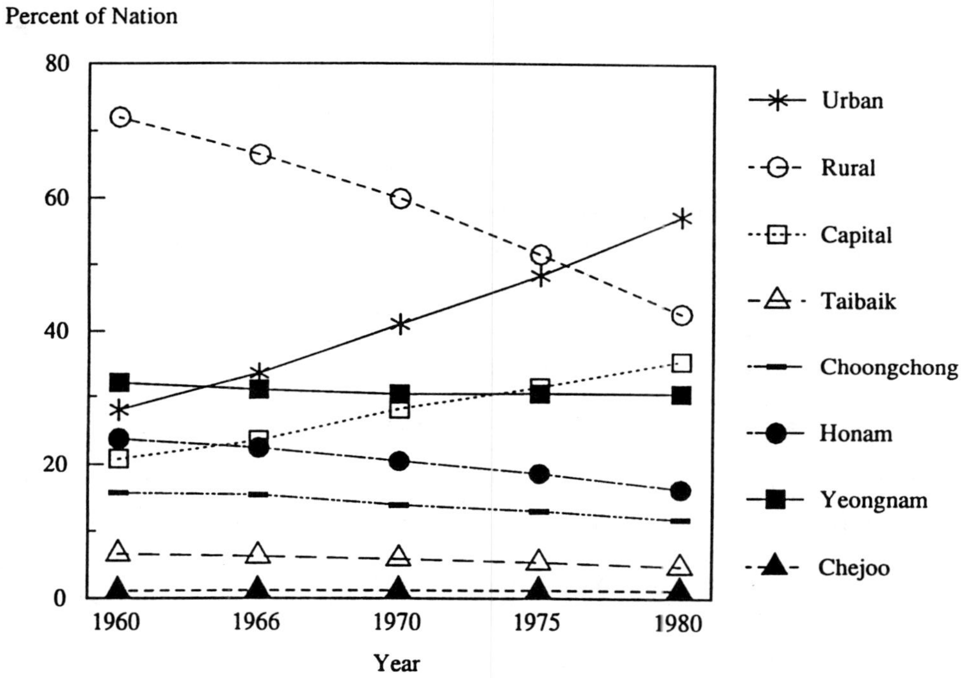

Figure 1.1: Population Trends by Region

and the Office of Secretariat of the President (Blue House)— have adopted seventeen different plans that deal with decentralization of socio-economic activities from Seoul between 1964 and 1981 (Table 1.3). The common aim of all seventeen plans was decentralization of socio-economic activities from Seoul, Pusan, and Daegu. The implementation of these plans, however, has always centered on Seoul, with the following goals:

1. Restriction on the new location of establishments of factories and universities in Seoul.

2. Relocation of government offices (particularly research institutes) to regions other than the capital region.

3. Construction of new satellite cities around Seoul.

4. Construction of industrial complexes in other regions.

5. Tax abatement for industries relocating from the capital region.

6. Reassignment of administrative authorities from the central government to regional governments.

Despite persistent efforts by the government, the level of concentration of economic activities in Seoul seems to be ever increasing. In fact, inconsistencies in the government's efforts toward balanced growth can be seen in the amount and share of investment on infrastructure development (Table 1.4). Seoul received more than fifty percent of the nation's investment in road construction in the past decade. About twenty percent went to the construction of roads in Pusan and the remainder to the rest of the country's roads.

Statistics in water supply shows a similar pattern. More than forty percent of the investment for the construction of a water supply system in Korea was spent in Seoul. About ten percent was spent in Pusan and the remaining fifty percent on the rest of the country for the past ten years or so. Investments on other social infrastructure developments, such as sewage improvement, power, and telephone, have been spent mainly in Seoul and Pusan. Figure 1.2 contrasts the timing of the government's major strategies for decentralization and share of investment for infrastructure in Seoul. The inconsistency between the decentralization strategies and infrastructure investment policy is quite apparent. This inconsistency, of all issues, has created the most concern.

Date Adopted	Prepared	Title	Remarks
Sept. 22, 1964	MOC[1]	Population Decentralization Plan	Approved by the Cabinet Council
Dec. 1969	MWP[2]	Population Decentralization Plan from the Capital Region	
Apr. 3, 1970	MOC	Strategies for Decentralizing Population from the Capital Region	Approved by the Cabinet Council
Oct. 27, 1971	MOC	Alternative Development Plans for the Capital Region	Approved by the Cabinet Council
1972	City of Seoul	Study of Decentralization of Population	
Oct. 1972	MOC/Blue House	Strategies for Population Decentralization from Metropolitan Areas	
Jan. 17, 1973	MOC	Comprehensive Plan for the Capital Region	
Mar. 9, 1973	EPM[3]	Population Decentralization Strategies from Metropolitan Areas	
	Blue House	Land Use Policies & Urban Planning for Population Decentralization	
Jan. 1975	MOC	Population Decentralization Strategies	
1975	MOC	Strategies for Urban & Regional Developments	
Aug. 1975	City of Seoul	Plans for Population Decentralization	
Oct. 1975	EPB	Strategies for Population Decentralization from Seoul	
Oct. 2, 1976	MOC	Plan for Banweol New Town: A Satellite City of Seoul	Approved by the President
Dec. 1977	MWP	Comprehensive Plan for Population Redistribution from the Capital Region: 1977-1986	Approved by the Cabinet Council
Dec. 1979	KRIHS[4]	Alternative Plan for the Capital Region	
Dec. 1981	KRIHS	Comprehensive Plan for the Capital Region	

Note : 1. Ministry of Construction
2. Ministry Without Portfolio
3. Economic Planning Board
4. Korea Research Institute for Human Settlements

Table 1.3: The Government's Decentralization Plans for Socio-Economic Activities in Korea

1. Road (expenditure for public works - percent share)				
Year	Seoul	Pusan	Rest	Total (1000 Won)
1972	46.29	21.30	32.41	8,979,126
1974	16.01	21.65	62.35	8,952,491
1976/1977	47.58	10.87	41.55	21,146,019
1979	75.15	10.52	14.33	154,412,565
1980	64.29	14.48	21.13	196,381,392
1981	52.62	19.80	27.58	210,306,324
1982	33.49	14.68	51.83	102,093,420
1983	36.82	11.69	51.49	156,270,399

2. Water Supply (expenditure for public works - percent share)				
Year	Seoul	Pusan	Rest	Total (1000 Won)
1972	43.20	4.10	52.70	12,942,265
1974	0.03	10.12	89.85	5,175,928
1976/1977	40.17	7.54	52.29	46,372,593
1979	40.82	16.23	42.95	52,635,102
1980	54.07	14.06	31.87	100,186,014
1981	49.24	14.39	36.37	142,263,309
1982	28.86	24.32	46.82	118,478,218
1983	20.89	25.58	53.53	151,077,724

3. Sewage (expenditure for public works - percent share)				
Year	Seoul	Pusan	Rest	Total (1000 Won)
1972	63.55	4.17	32.38	5,583,249
1974	45.25	2.05	52.50	2,653,909
1976/1977	60.88	6.81	32.31	8,635,899
1979	39.75	11.52	48.73	14,300,939
1980	37.89	16.59	45.52	15,140,233
1981	63.24	4.52	32.24	63,293,250
1982	26.91	5.75	67.34	25,499,372
1983	23.71	6.31	69.98	43,584,848

Source : Korea Municipal Yearbook, Corresponding Years

Table 1.4: Investment on Infrastructure Development in Korea

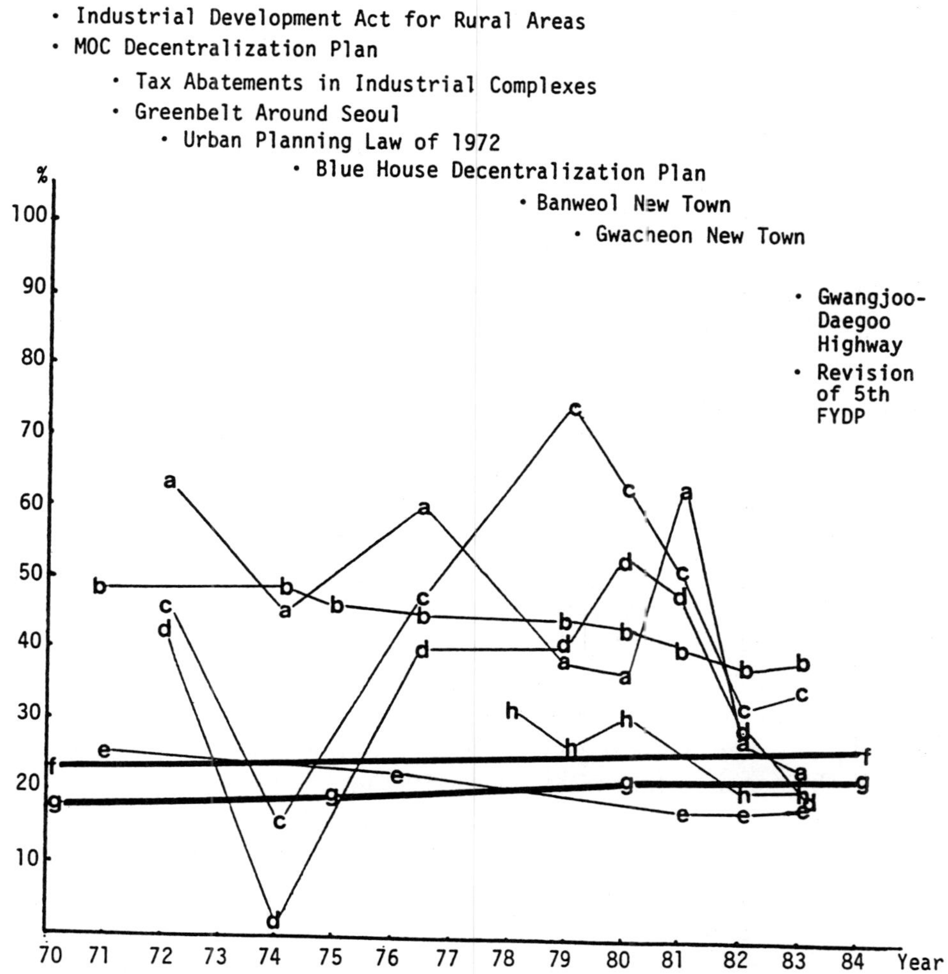

Note: The letters in the Figure correspond to investment share (%) in the following categories:

a. Sewage
b. Telephone
c. Road
d. Water Supply
e. Power
f. Manufacturing Firms
g. Population
h. Public Office & Hospital

Figure 1.2: Major Decentralization Strategies in Contrast to Infrastructure Investments in Seoul

1.3.2 Unbalanced Regional Growth

The level of concentration of economic activities in Seoul has persistently increased in the face of government policies against concentration (Table 1.2). Kim (1984)[133] pointed out three major reasons for the failure of government efforts to achieve a balanced distribution of socio-economic activities in Korea. The first and foremost is the lack of provision for improving the welfare of the people in rural areas. The second is the government's emphasis on the dispersion of the physical location of manufacturing complexes rather than the tertiary activity, which has been the fastest growing sector in Seoul. The final cause of the failure is the government's providing strong implicit incentives, such as taxes and export subsidies, that favor industries locating in Seoul or nearby.

1.3.3 Transportation Planning Issues

National Viewpoint

Due to Korea's heavy dependence on exports for its economic growth, it matters a great deal how much and where transportation services are provided. During the course of planning and implementing national transportation development programs, programs as well as investment priorities have been constantly revised due to changes in external and internal economic climates. One year after the adoption of the fifth FYDP in 1982, for example, the Economic Planning Board of Korea in 1983 announced the revision of plans in response to changes in the external and domestic economic variables. Three important factors were cited for the revision: the declining price of crude oil, the unforeseen delay in the world economic recovery, and improvement in the national balance of payments (Suh 1983)[240].

In addition, the agricultural sector has lagged far behind the manufacturing sector. While the decline in crude oil prices helped to improve the nation's balance of payment, the subsidy needed for stabilizing rice and fertilizer prices made payments to the farming sector one of the major causes of budget deficits. Revised transportation development programs now include raising the opportunity for farmers' real income through expanded infrastructure, particularly improvements of rural roads.

Regional imbalance of growth also poses unique transportation problems in Korea, because the movement of vehicles and fleets among regions forces a majority of vehicles to return to the origin of shipment empty. This implies that backhaul trips become crucial for representing true movement of vehicles and fleets among regions in Korea. Thus, proper representation of vehicle size in transportation planning processes is an important issue, because it affects operating costs, energy consumption, congestion, pollution, maintenance cost of networks and, ultimately, the planning and design network system.

Urban Viewpoint

If the current population and urbanization trend continues, it is not difficult to foresee that Korea's population will grow to fifty million and eighty percent will live in urban areas by the year 2000. The alarming implication of this forecast is that, within the next ten years, either new cities will have to be built or existing cities will have to be expanded to accommodate sixteen million new urban dwellers.

With this striking trend, which may soon become reality, regional scientists and transportation analysts in Korea must confront the issue of efficient resource allocation between urban land uses and transportation developments. With a limited amount of land resources, transportation analysts must deal with ways to reduce congestion. At the same time, the improvement should not be a cause for additional influx of migrants to metropolitan areas.

1.3.4 The Greenbelt Policy

The City Planning Law of 1971 empowered the government to designate greenbelts around major urban areas. Development in the greenbelt areas was prohibited, except for special projects approved by the authorities and certain other types of development.

The first greenbelt of 166.8 km^2 was established around Seoul and later expanded to cover twelve major urban areas covering about 5,397 km^2 in the suburban areas of major cities. The main purpose of the greenbelt policy was to restrict the growth of Seoul and to spread additional urban growth to medium-sized urban areas nearby. This policy appears to be strictly enforced, and no development has actually taken place (Kim 1984)[133].

The greenbelt policy, however, failed to stem the population growth in Seoul, only dispersing it to the adjacent Gyunggi area. Even worse is that the greenbelt policy was a major cause of the rapid increases in land and housing prices and high urban densities during the late 1970s.

1.4 Concluding Remarks

Based on the discussion of the issues related to the spatial distribution activities in Korea in preceding sections, the rest of chapters of the book are devoted to answering many important related questions, including the following:

1. Are there ways to model the proper interaction between transportation and the rest of the economic systems? Such a model will not only be necessary for developing programs for national development, but also will be needed for identifying and revising priorities in implementing transportation development programs once external and domestic economic indicators are changed.

2. What are the alternative ways to identify "optimum investments" for transportation systems in Korea? How do we identify those?

3. What are the optimum levels of governments' intervention in mixed economic systems? How do we model them? How can we properly represent both private and public sectors' objectives in modeling national development?

Chapter 2

The Transportation Systems in Korea

2.1 National Development Plans and the Transportation Sector

During the five consecutive FYDP periods (1962–1986), the government of Korea invested heavily in infrastructure development, particularly on transportation.[1] Almost half of the total budget was spent on transportation during the first FYDP period (1962–1966). The government invested heavily in railroad systems to provide greater accessibility to and within the seven special planning districts established during the first FYDP (1962- -1966) period.

Two expressways were opened during the second FYDP (1967–1971): the 29.5-km Seoul-Incheon expressway in 1968 and the 428-km Seoul-Pusan expressway in 1970. While highway expansion was still a dominant part of transportation investment during the third FYDP (1972–1976), the overall thrust of the national planning effort was the construction of heavy and chemical industry complexes. During the fourth FYDP (1977–1981), development of an integrated transportation system received high priority in the transportation sector. Subways in Seoul and Pusan were initiated during the period and investment in highways was still dominant (Table 2.1). The subway networks in Seoul and Pusan were almost completed during the fifth

[1]Portion of this chapter draws upon Sunduck Suh's doctoral dissertation, *Implementation and Evaluation of Nonlinear Bilevel Programming Model of Equilibrium Network Design Problem*, University of Illinois at Urbana-Champaign, Urbana, IL. January, 1989.

FYDP (1982–1986), while marine and port transport received significant investment capital. The whole development of the transportation system in Korea is summarized in Table 2.1.

2.2 Transportation System

The development of the transportation system in Korea has had a very close relationship with the growth of the national economy. This is depicted in Figure 2.1, which shows the close relationship between Gross National Product (GNP) and transport value added (TVA). Further, the transportation sector makes up a significant proportion of the Gross Domestic Product (GDP) (Table 2.2).

The proportion of the transport sector as a percentage of GDP remained almost constant from 5.7 percent to 6.0 percent from 1980 to 1985 (Table 2.2). Among all modes of transportation, highways demonstrated the highest annual increase in TVA during the 1980–1985 period (27.3 percent). Railways, on the other hand, showed the smallest annual increase (1.1 percent), during the same period. The highway system has dominated domestic passenger transportation (Figure 2.2).

Highway systems also play a dominant role in freight transport. The shares for the different modes available for freight movements are shown in Figure 2.3.

2.3 The Highway System

The dominant role of the highway system in transporting passengers and freight is best demonstrated by the rapid increase in the number of motor vehicles (Table 2.3). As can be seen in Table 2.4, about 600 km of national roads and about 550 km of provincial roads were paved between 1984 and 1985. These paving activities usually are accompanied by improvements in the geometric design of roads and increasing thoroughfare capacity of roads. No doubt this trend of rapid improvement will continue in the future, because a large portion of provincial roads are unpaved (seventy percent).

The highway network used in this book is based on the road conditions in 1985 for fifteen and twenty-five origin/destination (O/D) zonal systems. The size of the networks are 221 nodes and 706 links for twenty-five O/D zonal systems, and sixty-three nodes and 190 links for fifteen O/D zonal

Sector	First[1] '62-'66	Second[1] '67-'71	Third[2] '72-'76	Fourth[2] '77-'81	Fifth[2] '82-'86
Total	1,048.0 (100)	3,045.0 (100)	11,404.0 (100)	24,152.1 (100)	72,100.0 (100)
Agriculture Forestry and Fisheries	106.7 (10.2)	214.9 (7.1)	1,056.0 (9.3)	1,802.6 (7.5)	6,300 (8.7)
Mining and Industry	240.4 (22.0)	601.0 (19.7)	2,649.0 (23.2)	5,510.5 (22.8)	16,700.0 (23.2)
Social Infrastructure and Service	598.9 (57.2)	2,011.0 (66.2)	4,697.0 (58.7)	16,839.0 (69.7)	39,600.0 (55.0)
Others	102.0 (9.7)	218.1 (7.2)	1,002.0 (8.8)	- -	9,500.0 (13.1)
Transportation	102.9 (9.8) (100)	472.6 (7.2) (100)	1,796.0 (8.8) (100)	1,873.7 (7.8) (100)	9939.7 (14.1) (100)
Rail	49.8 (48.4)	78.0 (62.3)	274.0 (46.9)	307.2 (54.4)	1,330.8 (37.4)
Highways	31.5 (30.6)	294.2 (62.3)	843.0 (46.9)	1,019.2 (54.4)	3,718.8 (37.4)
Marines and Ports	19.5 (19.0)	92.2 (19.5)	497.0 (27.7)	270.9 (14.4)	2,655.1 (26.7)
Aviation	2.1 (2.0)	8.0 (1.7)	134.0 (7.5)	61.4 (3.3)	594.5 (6.0)
Subways			48.0 (2.6)	215.0 (11.5)	1,635.0 (16.5)

Note : Units are in billion Won
[1] 1970 Constant Won
[2] 1975 Constant Won

Source : Economic Planning Board

Table 2.1: Investment Amounts by Sector and Planning Period

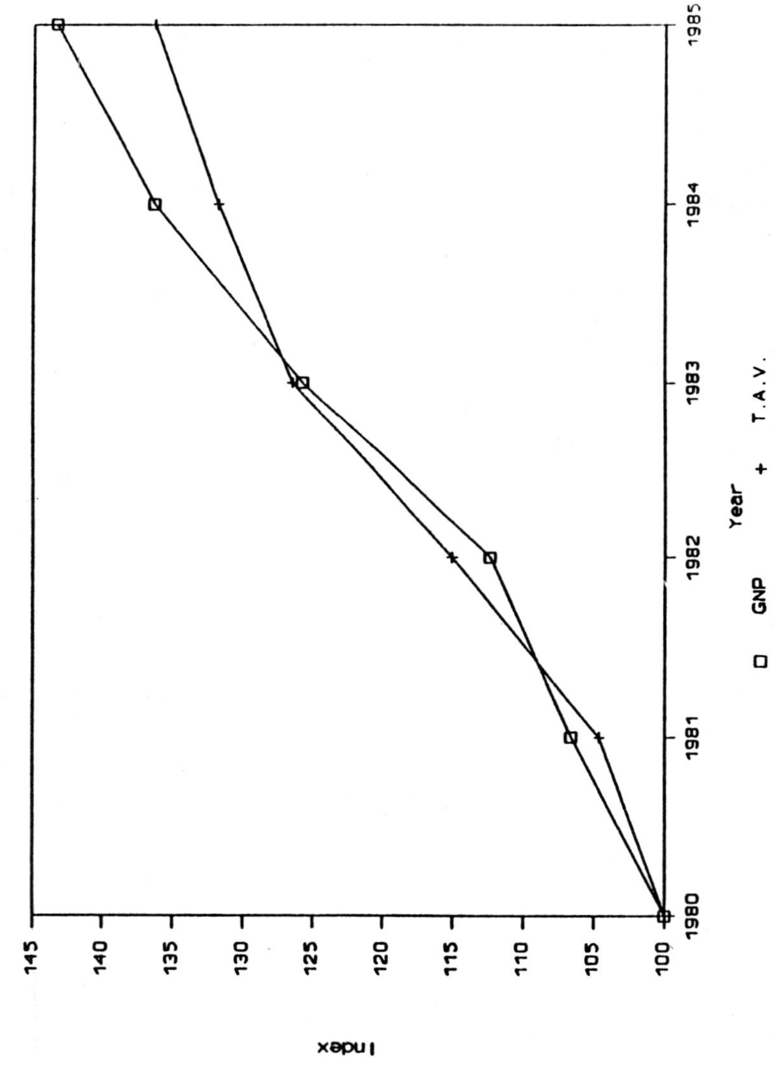

Figure 2.1: Growth of GNP and Traffic Value Added

Sector	1980	1981	1982	1983	1984	1985
G.D.P.	37914.9	40723.3	43035.7	47743.7	51872.5	54544.8
Agriculture Forestry and Fisheries	5524.7	6759.7	6980.6	7436.0	7453.2	7893.4
Mining and Industry	11734.3	12675.0	13185.3	14780.6	16930.4	17551.8
Construction Electricity Water	3936.0	3879.7	4476.4	5406.7	5959.8	6372.8
Service	13191.0	13735.5	14577.8	16171.9	17490.9	18558.6
Other	3528.9	3673.4	3815.6	3948.5	4038.2	4168.2
Transportation	2304.4 (6.1) (100)	2405.2 (5.9) (100)	2463.4 (5.7) (100)	2902.9 (6.1) (100)	3027.4 (5.8) (100)	3136.3 (5.7) (100)
Rail	189.9	185.3	160.2	168.6	173.4	190.7
Highways	1284.1	1313.6	1485.6	1668.2	1859.4	1959.1
Shipping	456.4	498.9	575.4	617.6	528.9	501.9
Aviation	166.9	183.0	195.0	210.9	225.3	235.9
Other	210.2	229.7	247.8	268.6	286.7	299.3
Storage	59.4	67.9	77.1	86.6	88.1	86.3

Note : Units are in billion Won (1985 Constant Won).
Numbers within parentheses under transportation indicate the share (in percentages) of the transportation sector in GDP.
Source : The Bank of Korea

Table 2.2: Gross Domestic Product and Transportation Sector

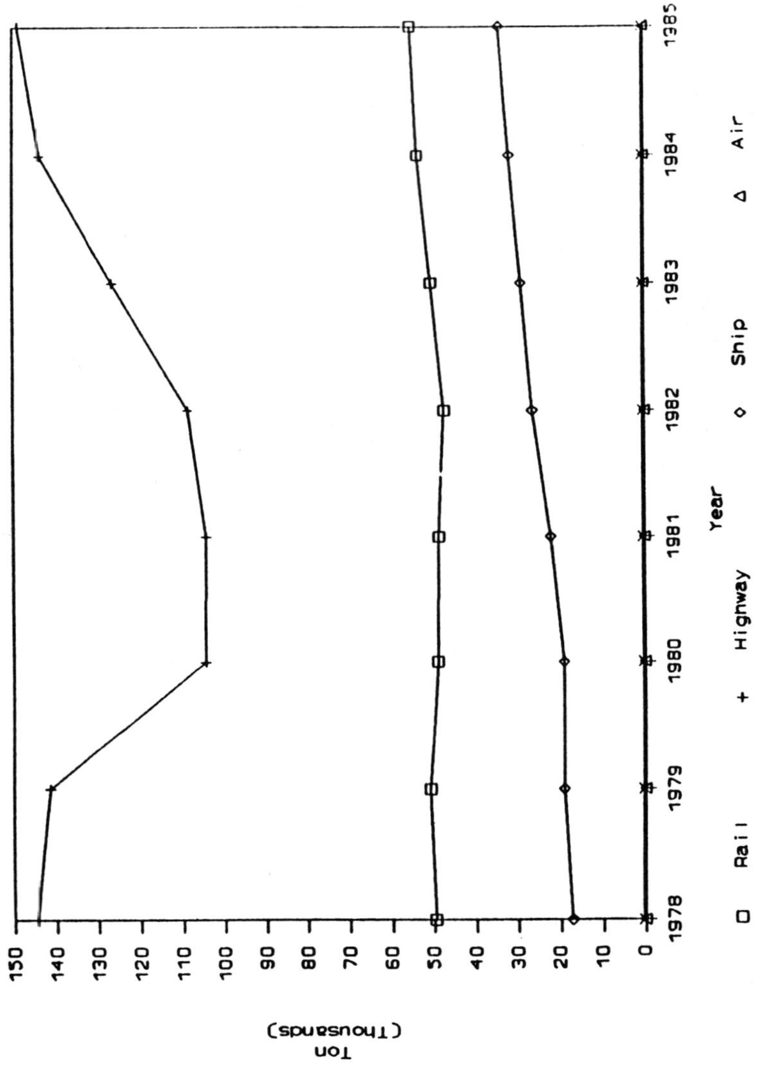

Figure 2.2: Domestic Passenger Traffic

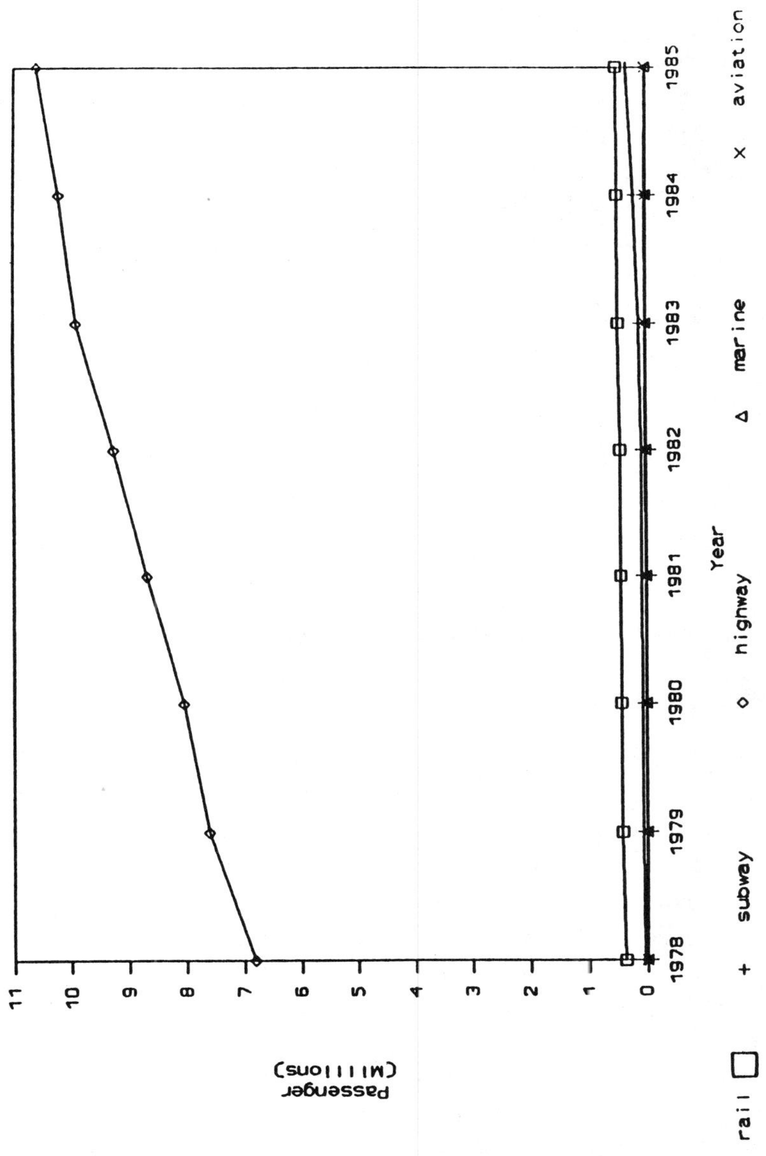

Figure 2.3: Domestic Freight Traffic

Year	Population (× 1000)	Vehicle	Passenger[1] (× 1000)	Freight[1] (× 1000 ton)	Road (Km)
1978	36969	125660	6807990	144599	45954.7
1979	37534	146113	7609676	141512	46333.0
1980	38124	149710	8039005	104525	46950.9
1981	38723	157243	8683129	104255	50336.6
1982	39331	163650	9259252	108575	53935.5
1983	39951	182645	3901324	126402	54599.6
1984	40578	201608	10200618	143628	51003.7
1985	41056	211740	10601046	148699	52264.1

Note: [1] for highway only
Source: Ministry of Transportation, 1986 [5]

Table 2.3: Trends of Major Transportation Related Indices

systems. Figures 2.4 and 2.5 show the highway networks and O/D zones used in this book.

2.4 Highway Investments in Korea

2.4.1 Introduction

An investment function gives the relationship between the capacity increases and the cost incurred with the link improvements. The investment function and the budget for road improvements actually helps decision makers determine the amount of capacity increase in a network design model.

Abdulaal and LeBlanc (1979)[1] reported that small improvements over many links were induced with a convex investment function. If a concave function is used, large improvements over a small number of links are generated. This improvement specification is realistic in that it represents the economies of scale.

Despite the importance of the investment function, not much research has been done on this issue. Steenbrink (1974)[6] reported several piece-wise linear investment functions for roads and intersections of different types for the Dutch environment. Recently, Ben-Ayed (1988)[2] used a piece-wise linear investment function that accounts only for the surface treatment of a two-lane highway for Tunisia.

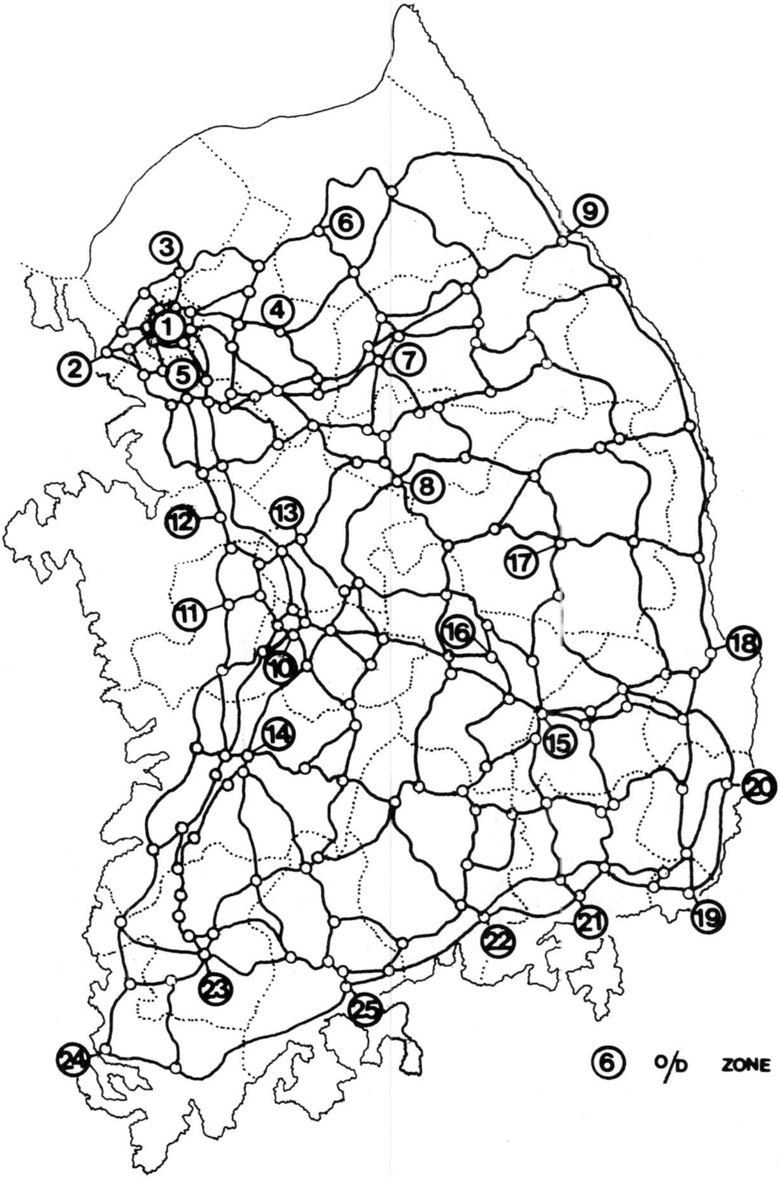

Figure 2.4: Highway Network with 25 O/D Zones

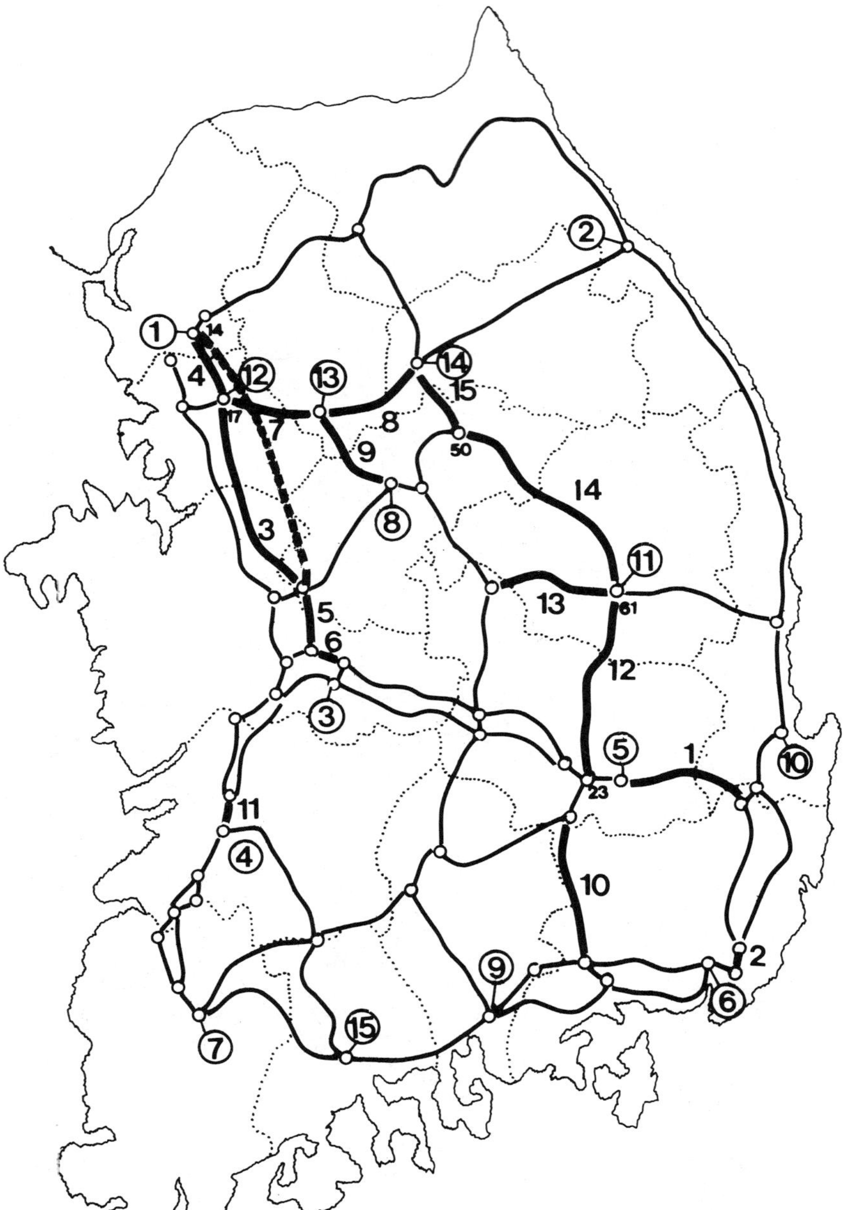

Figure 2.5: Highway Network with 15 O/D Zones

Year	National Road		Provincial Road		Total (Km)		
	Paved	Gravel	Paved	Gravel	Paved	Gravel	Total
1975	3619.8	4600.6	740.8	9769.0	9999.8	31295.4	44905.4
1976	3746.0	4474.4	797.0	9768.0	10912.2	31031.3	45513.6
1977	4223.6	3996.8	907.8	9691.4	12096.8	30425.5	45663.8
1978	4974.5	3245.9	998.1	9597.2	13543.5	29009.0	45954.7
1979	4999.8	3220.6	1168.8	9586.4	14277.7	29144.8	46333.0
1980	5546.8	2673.7	1384.8	9403.4	15599.0	28337.6	46950.9
1981	6774.3	5401.5	1102.6	8529.4	17178.9	29839.9	50336.6
1982	7397.9	4754.7	1285.3	8415.0	19282.7	29085.1	53935.5
1983	8191.8	3962.3	1826.9	7959.9	21278.8	27936.9	54599.6
1984	8427.4	3739.7	2404.3	7401.6	23664.3	26403.5	51003.7
1985	9017.7	3441.1	2966.6	6941.9	26072.0	25293.9	52264.1

Source: Ministry of Construction

Table 2.4: Length of Paved and Unpaved National and Provincial Roads (unit: Km)

The purpose of this section is to derive an investment function that is suitable for Korean highway improvement. Particularly, this study involves the paving of two-lane unpaved roads and the upgrading of two-lane roads to four-lane roads. The data used are described in the next subsection, followed by the numerical analysis for developing an investment function.

2.4.2 Data

During the time encompassed by the five consecutive FYDPs, Korea has implemented a series of highway improvement projects. The funds for a number of recent projects came from the Asian Development Bank (ADB) and the International Bank for Reconstruction and Development (IBRD). Specifically, there have been three ADB- and five IBRD-funded road improvement projects. The feasibility study of a fourth ADB road improvement project has been completed recently; and the data used in this section for the development of the investment function are from the Feasibility Study Report for the ADB Fourth Road Improvement Project (MOC 1987)[4]. The project's objectives include studies on the technical and economic feasibility of improvements on 640 kilometers of national highway. More specifically, it studied the feasibility of upgrading and paving a total of 466.5 kilometers of two-lane road segments and of widening 173.6 kilometers of two-lane road to four-lane road. These road segments are scattered around Korea and represent diverse land use environments and terrain characteristics, from flat to mountainous. Flexible pavement is used in the paving of a two-lane unpaved road, and rigid (concrete) pavement is utilized in the widening project of two-lane to four-lane paved road.

The average cost per kilometer of two-lane paving projects, including land cost, is 449 million Won[2] and 1,165 million Won for four-lane widening. The capacity increases are 5,000 vehicles per day for two-lane and 35,000 vehicles per day for four-lane (for more details, see MOC 1987 [4]). The costs of adding one capacity unit per kilometer, therefore, are 0.0898 million Won and 0.0333 million Won for two-lane and four-lane improvement, respectively. These decreasing unit cost figures strongly suggest the existence of economies of scale and thus a concave investment function. This will be discussed in detail in the next subsection.

For construction cost, the report details cost breakdown among direct construction cost, land cost, earthwork, pavement, drainage, structure, and miscellaneous cost. The economic cost is used for investment analysis. It is observed that pavement accounts for fifty percent of the total direct construction cost. The cost of materials such as gravel and sand for pavement work account for about fifty percent of the total pavement cost (MOC 1987)[4]. Therefore, the cost of materials is a very important factor in the total construction cost. Furthermore, because the cost of materials is heavily dependent on the hauling distance, the distance to the source of materials is included in the analysis. Capacity increases, distance to the source of materials, economic cost of construction and right-of-way, and terrain type (1=hilly, 2=mountainous, 3=flat in Table 2.5) are chosen as the decision variables in the analysis of investment on highways on Korea.

2.4.3 A Highway Investment Function in Korea

Based on the data previously described, the four decision variables are tabulated (Table 2.5).

A regression analysis is applied to find the relationship among capacity increase, construction cost, distance to the source of material, and terrain type using the SPSS package[3]. The relationship between total cost and these factors is also examined. Table 2.6 shows the correlation matrix among

[2] Korean currency, US dollar=740 Won as of September, 1988.

Link	Cap. Inc. veh/day	Dist. to Mat. (Km)	Const. Cost Mil. Won (Km)	Land	Total	Terrain
1	37800	13.5	569	438	1007	1
2	35800	13.5	940	90	1030	2
3	39700	19.5	467	120	587	1
4	42800	19.5	618	313	931	2
5	35400	19.5	536	174	710	1
6	37300	83	1079	510	1589	1
7	39500	15.2	705	216	921	1
8	38500	15.2	793	170	963	1
9	39400	18	780	117	897	1
10	38900	18	920	203	1123	1
11	36500	45	708	86	794	1
12	38000	25	1157	111	1268	1
13	37000	60	1221	76	1297	1
14	32900	60	1404	646	2050	1
15	10250	34	496	104	600	1
16	4250	34	464	76	540	2
17	4250	34	395	78	473	2
18	1900	11	535	20	555	3
19	1900	1	270	0.35	270.35	3
20	4250	1	250	0	250	2
21	1900	13.6	465	17	482	3
22	4250	26.2	609	15	624	2
23	4250	26.2	657	12	669	2
24	4250	0.4	131	0	131	2
25	1950	0.4	258	0	258	3
26	4250	11	299	22	321	2
27	1900	11	431	32	463	3
28	4250	27.9	478	25	503	2
29	4250	27.9	392	24	416	2
30	10250	13	396	61	457	1
31	4250	13	314	39	353	2
32	1900	28	372	47	419	3
33	4250	23.2	386	46	432	2
34	10250	23.2	343	29	372	1
35	10250	22	362	31	393	1
36	4250	22	362	31	393	1
37	4250	68	459	51	510	2
38	10250	54	367	59	426	1
39	10250	54	367	59	426	1
40	10250	54	368	57	425	1
41	10250	22.2	353	31	384	1
42	10250	22.2	295	61	356	1
43	1900	22.2	293	22	315	3
44	1900	22.2	413	21	434	3
45	4250	40	375	34	409	2
46	10250	40	338	34	372	1
48	4250	27	362	29	391	2
49	10250	27	301	45	356	3
50	1900	4	458	45	503	3
51	1900	4	419	27	627	3
52	1900	4	600	27	479	3
53	1900	4	452	27	479	3
54	1900	4	452	27	479	3
55	4250	18	338	40	378	2
56	4250	18	343	20	363	2

Table 2.5: Capacity versus Cost Data from the 4th ADB project

	Capacity	Construction	Distance	Land
Construction	0.72616			
Distance	0.23568	0.39144		
Land	0.68260	0.66776	0.38316	
Total	0.77038	0.96507	0.42063	0.83946

Table 2.6: Correlation Matrix between Variables

these factors.

Based on the preliminary analysis of Section 2.4.2, which suggests a concave investment function, the following nonlinear equation was assumed to be the basic functional type.

$$\mathrm{COST} = aX^bS^c \tag{2.1}$$

COST : construction cost (million won/km)

X : capacity increase (vehicles/day)

S : distance to the source of material (km)

To account for three different terrain types, two dummy variables, D_1 and D_2, are introduced.

	D_1	D_2
Hilly	10	1
Mountainous	1	10
Flat	1	1

Including these dummy variables, the final equation becomes:

$$\mathrm{COST} = aX^bS^cD_1^eD_2^f \tag{2.2}$$

Using the SPSS package, the regression equations for the construction cost per kilometer and total cost per kilometer, including right-of-way, are given as follows:

$$\begin{aligned} \text{CONSTRUCTION COST} &= 2.56X^{0.48}S^{0.18}D_1^{0.18}D_2^{0.48} \quad (2.3)\\ R^2 &= 0.6222 \\ F &= 27 \quad \text{Significance} = 0.00 \\ \text{TOTAL COST} &= 1.26X^{0.57}S^{0.20}D_1^{0.19}D_2^{0.52} \quad (2.4)\\ R^2 &= 0.75958 \\ F &= 40.28 \quad \text{Significance} = 0.00. \end{aligned}$$

Both equations and their coefficients are significant at ninety-five percent confidence intervals and both equations give almost the same predicted cost figures up to four-lane expansion. Because of its better R^2 value, however, total cost is utilized in the forthcoming analysis.

Part II

LINEAR PROGRAMMING MODELS AND APPLICATIONS

Chapter 3

A National Transportation Development Planning Model

3.1 Introduction

How can we model the proper interaction between transportation and the other economic systems?[1] As described before, such a model will not only be necessary for developing programs for national development, but also will be needed in revising priorities and implementing program changes. National transportation development planning models fall into two categories: positive and normative, like other models for both mixed and planned economies. The usual view in mixed economic systems is that some goods and services are produced by the private sector and some by the public sector.

Normative concepts are much more primitive in urban, regional, and national development planning models than elsewhere in economics. Most normative models employ a standard paradigm: a market resource allocation is derived from a model, the market failure is shown, and certain public service needs, such as transportation, are derived from the solution despite the fact that the interaction between the public and private sectors does not appear in the model and that the model contains no welfare criterion (Mills 1975)[179]. This common approach misses the essential normative issue: the interaction between the public and private sectors.

---

[1]Portions of this chapter was written with Jong Gie Kim, and was published in *The Annals of Regional Science*, March 1985, 19:18–36.

Few would contend that private firms could or should construct an efficient transportation system in a mixed economic system. This is because the entire transportation system must be integrated and coordinated, and a firm with the power to do so would have considerable monopoly control. Thus, public sector intervention is required to construct and maintain an efficient urban, regional, and national transportation system. An important question arises here: How much public intervention should there be?

The question becomes more critical in developing economies, where goals are often established but are difficult to achieve because of limited or scarce resources and inadequate transportation systems. The issue here is not only the appropriate amount of public intervention but also the type and location of services that government should provide. Transportation is an important public service in regional and national economic activities. All private activities require the movement of goods and people. Thus, both the location and means of production and the origins and destinations of trips are affected by the provision of transportation facilities. At the same time, the location of private activities affects the demands placed on the transportation system. It matters a great deal how much and where transportation is provided in the multiregional economic system.

Multiregional planning and model building in a mixed economy is complex. A model that is to be useful for national and/or multiregional planning purposes must be capable of providing not only a detailed spatial representation of the regional economy, but must also induce the competitive profit-seeking firms to allocate resources efficiently. The following section introduces such a model.

3.2 National Transportation Planning Models

Since the 1960s, developing countries have made considerable progress in raising productivity and real incomes and in improving infrastructure such as transportation networks. The progress often has been a direct result of successful implementation of national development programs, such as FY-DPs. At the same time, several regional and interregional transportation planning models have been built and have become bases for the development of future plans. An excellent review of multiregional economic development models can be found in Rietveld (1981)[212] and Friesz, Tobin, and Harker (1983)[104]. The latter authors presented a topology of predictive freight models with respect to routing and model choice characteristics. Models

	Selective Characteristics	HB	CACI	PRINCETON	FNEM	RDRI/KIST
1.	Multimode	Y	Y	N	Y	Y
2.	Multicommodity	Y	Y	Y	Y	Y
3.	Congestion	N	Y	N	Y	N
4.	Sequential Macro-economic and Network Models	Y	Y	Y	Y	Y
5.	Transportation Demand	Y	F	F	Y	Y
6.	International Linkage	Y	N	N	N	Y
7.	Vehicle Size Model	N	N	Y	Y	N
8.	Backhaul Model	N	N	N	N	Y
9.	Network Equilibrium	O	O	O	W	O
10.	Shipper/Carrier Interaction	N	N	Y	Y	N

Note:

- HB = Harvard-Brookings
- Y = Yes
- N = No
- F = Fixed
- W = Wadrop's Principle: see p. 65 for detailed descriptions
- O = Other

Table 3.1: Comparison of Selective Characteristics of Multiregional Multi-commodity Transport Planning Models

reviewed by these authors are Harvard-Brookings (1966[213], 1971[151]), Bronzini (1980)[48], CACI (1980)[53], Peterson and Fullerton (1975)[200], Lansdowne (1981)[156], the Princeton model by Kornhauser et al.(1979)[149] and the Freight Network Equilibrium Model (FNEM) by Friesz, Gottfried, and Morlok (1981)[101]. Several models for transport system investment problems can be found in Bergendahl (1969)[30], Barber (1975[14], 1977[15]) and Ahmed et al. (1976)[4].

Based on the topology developed by Freisz, Tobin, and Harker (1983)[104], Table 3.1 presents a brief summary of characteristics of selected models, including the one developed by RDRI/KIST (1980)[139], the outline of which shall be discussed in this section.

In 1979, the Regional Development Research Institute of the Korea Institute of Science and Technology (RDRI/KIST) was commissioned by the government of Korea to recommend transportation improvement programs to be implemented during the fifth FYDP period, 1982–1986.

The RDRI/KIST model consists of three sequential components: (1) Trip Generation and Network Design Model, (2) Mode Choice Model, and (3) Vehicle Size Model.

3.2.1 Trip Generation and Network Design Model

This model produces (1) interzonal commodity flows that are functions of the nation's import/export targets, (2) location and amounts of production as well as of consumption for each commodity, and (3) amounts of import/export goods that are to be handled by each port. Simultaneously, the model produces necessary capacity increases between pairs of regions for each type of network system, i.e., unpaved roads, two-lane highways, four-lane highways, and railroad systems. Based on additional requirements of capacity increases between pairs of regions, alternative networks are generated.

The model is as follows:

$$P3.1)\min \quad C = \sum_i \sum_r p_r^i X_r^i + \sum_k \sum_{ij} (a_k K_k^{ij} + b_k L_k^{ij}) + \sum_r \sum_i (d_r^i F_r^i + e_r^i G_r^i) \tag{3.1}$$

$$\text{s.t.} \quad \sum_{i \in e} E_r^i \geq E_r, \quad \sum_{i \in e} I_r^i \geq I_r \qquad \forall \ r \tag{3.2}$$

$$E_r^i + I_r^i \leq F_r^i + G_r^i \qquad \forall i, r \tag{3.3}$$

$$\sum_r F_r^i + \sum_r G_r^i \leq Q^i \qquad \forall i \tag{3.4}$$

$$I_r^i + \sum_j \sum_k X_{rk}^{ji} + X_r^i = E_r^i + \sum_i \sum_k X_{rk}^{ij} + \sum_q a_{rq} X_q^i + E_r^i \qquad \forall \ i, r \tag{3.5}$$

$$\sum_r X_{rk}^{ij} \leq K_k^{ij} + L_k^{ij} \qquad \forall \ i, j, k \tag{3.6}$$

$$E_r^i, I_r^i, X_r^i, X_{rk}^{ij}, F_r^i, L_k^{ij}, K_k^{ij} \geq 0 \qquad \forall i, j, k, r \tag{3.7}$$

where the endogenous variables are as follows:

E_r^i = amount of commodity r to be exported from port i

I_r^i = amount of commodity r to be imported through port i

X_r^i = amount of commodity r to be produced at region i

X_{rk}^{ij} = amount of r shipped from region i to region j using transportation system k

G_r^i = optimum expansion of port i for handling r

F_r^i = amounts of commodity r that can be handled within existing capacity of port i

L_k^{ij} = optimum expansion of transportation system k between regions of i and j

K_k^{ij} = amounts of commodity shipped using existing transportation system k between regions i and j.

Exogenous variables and corresponding values are as follows:

E_r = export target amounts of commodity for a given year

I_r = export target amounts of commodity for a given year

p_i^r = cost of producing unit amounts of commodity r at region i

a_k = user cost for shipping unit amount of commodity between regions i and j using system k. Unit user cost, $W/ton - km$ (W = Won, the Korean currency), is estimated as

- Unpaved Road (k=1) W 26.8/ton-km
- Paved Highway (k=2) W 15.6/ton-km
- Railroad (k=3) W 14.0/ton-km

b_k = cost for expanding unit distance for system k. This cost has been estimated as

- Paved (k=2) W 24,000/lane-km
- Railroad (k=3) W 56,000/track-km

a_{qr} = input amounts of commodity q required to produce one unit of commodity r. The input-output table of 60 × 60 sectors developed by the Bank of Korea (BOK) in 1975 was reduced to a 5 × 5 matrix for the study (Table 3.2).

0	1	2	3	4	5
1	0.09584	0.05717	0.09825	0.00760	0.00086
2	0.00145	0.00297	0.08112	0.00587	0.14677
3	0.14596	0.14188	0.49831	0.21610	0.31155
4	0.01702	0.10247	0.05678	0.12868	0.11140
5	0.00034	0.00162	0.00727	0.03462	0.06236

Note: Sector 1 : Agricultural, fisheries, and forest products
2 : Coal and mining
3 : Manufacturing, except nonmetallic
4 : Service
5 : Cement and nonmetallic

Table 3.2: Input-Output Coefficients for a 5 × 5 Sector Matrix in Korea

The grouping of commodities is as follows:

- Sector 1(r=1): agricultural, fisheries, and forest products (sectors 1-6 of BOK table)
- Sector 2(r=2): coal and mining (sectors 7-9 of BOK table)
- Sector 3(r=3): manufacturing, except nonmetallic (sectors 10-33 and 35-44 of BOK table)
- Sector 4(r=4): service (sectors 45-50 of BOK table)
- Sector 5(r=5): cement and nonmetallic (sector 34 of BOK table)

d_r^i = handling cost (user cost) per unit amount of commodity r at port i (Table 3.3)

e_r^i = construction cost for expanding unit increase in port handling capacity of commodity r (Table 3.3)

Q_i = existing capacity of port i per year (see Table 3.3)

3.2.2 Mode Choice Model

No individual shipper's or carrier's behavior is explicitly represented in the mode choice model. However, an aggregate decision maker between the two

Region	Port	$Q(i)$ Existing Capacity (1,000MT)	d_i^r Unit User Costs (W1,000/1,000MT) r=2,5	r=1,3,4	e_i^r Unit Expansion Costs (W1,000/1,000MT) r=2,5	r=1,3,4
1	Incheon	21,000	1,630	1,930	2,130	3,196
2	Asan	-	-	-	1,500	2,630
3	Gunsan	2,500	920	1,220	1,420	2,590
4	Mogpo	800	920	1,220	1,420	2,590
5	Yeosu/Samil	15,000	730	1,030	1,230	2,690
6	Samcheonpo/Gunho	200	920	1,220	1,420	2,120
	Chungmu/Ogpo Jangseungpo/ Anjeong					
8	Masan/Jinhae	3,300	1,230	1,530	1,430	2,450
9	Busan	27,000	1,230	1,530	1,730	2,750
10	Ulsan	22,000	930	1,230	1,430	1,620
11	Pohang	22,000	920	1,220	1,420	2,620
12	Samcheog/ Bugpyeong	10,000	930	1,230	1,430	2,620
13	Gangreung /Mugho	6,500	630	930	1,130	3,420
14	Sogcho	600	500	800	1,000	2,320

Note: MT: Metric tons
W: Korean currency (Won)

Table 3.3: Existing Capacity, Unit User Costs, and Unit Expansion Costs for Ports

is assumed to choose the shipment model based on a logit function, as shown below:

$$p_k^{ij} = \exp(-\lambda c_k^{ij} / \sum_k \exp(-\lambda c_k^{ij}) \tag{3.8}$$

where

$$c_k^{ij} = a_k + \sum_l b_l x_{kl}^{ij}$$

c_k^{ij}	=	generalized cost between regions i and j using mode k
x_{kl}^{ij}	=	the cost in category l that shippers perceive using mode k between i and j
p_k^{ij}	=	probability of choosing mode k for shipping commodities between regions i and j
λ, a_k, b_l	=	coefficients to be estimated

The generalized cost of mode k between a given O/D pair was used to estimate the probability of choosing mode k (between trucks and railroads). Calibration results are presented in Table 3.4.

3.2.3 Vehicle Size model

The vehicle size models consist of three components: the truck size submodel, backhaul submodel, and empty rail movement submodel.

Truck Size Submodel

In the truck size submodel, it is assumed that the size of a truck for shipping freight between two regions is related to distance, terrain type (flat, hilly, mountainous, and urban), and road type (expressway, high standard paved road, low standard paved road, and unpaved road). Grouping distance between regions into four classes (over 15 tons, 9-15 tons, 5-9 tons, 2-5 tons), the current practice was observed through a field survey (Figure 3.1). The figure illustrates that the better road types there are, the larger-sized trucks are being operated. From this observation, a model was built that provides the maximum permissible distance between zones for each size-class of truck for each of the sixteen combinations of four road types and four terrain types.

The truck size submodel is:

Commodity Sector	PH = a+b×SCPTH PR = b×SCPTR	
	a	b
1	3.0076 (13.91)	0.0007 (16.54)
2	1.2598 (14.73)	0.0012 (17.20)
3	2.3608 (14.34)	0.0006 (17.20)
4	Not Estimated	
5	1.5611 (10.14)	0.0013 (132.23)

Notes:

- PH = Proportion shipped by trucks on highways when other mode is railroad
- PR = Proportion shipped by rail
- a, b = Calibration results for coefficients
- SCPTH = Total cost for shipping unit volume by truck on highways
- LCPTR = Total cost for shipping unit volume by rail
- For the classification of commodity sector, see Table 3.2
- t-values are in parentheses.

Table 3.4: Calibration Results for Mode Choice for Freight Shipment in Korea

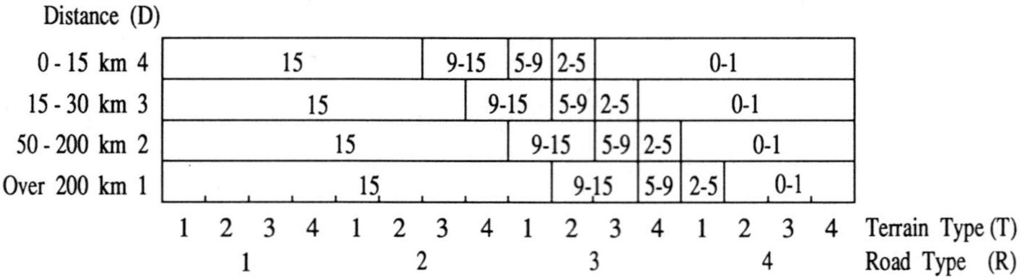

NOTE : 1. For road types and terrain types, see Table 3.5

2. Numbers in the figure represent truck size; 0 - 1 ton truck, 2 - 5 ton truck, 5 - 9 ton truck, 9 - 15 ton truck and over 15 ton truck.

Figure 3.1: Distance Traveled by Various-Sized Trucks on Different Terrains and Road Types in Korea

$$P_{DTRL} = 100/1 + \exp\{a + b_1 \cdot T + b_3[(R-1) \cdot 4 + L]\} \tag{3.9}$$

where

P_{DTRL} : Maximum permissible percentage of total journey in distance group D, for a vehicle in distance group D, for a vehicle in size-class T, traveling a road type R over terrain type L (Table 3.5). Calibration results are

- $a = -4.78868$
- $b_1 = -0.630425$
- $b_2 = -0.763125$
- $b_3 = 0.536299$

The equation (3.9) is applied in the following way. A score of P_{DTRL} is estimated for each link between regional pairs based on the terrain and road types and on distance. For a total journey between regional pairs, the average link scores, weighted by distance on each road type, was calculated. The largest-sized truck that gives a score over ninety-five percent is assigned. The ninety-five percent is chosen because it gives, for a selected sample of journeys, truck sizes that correspond to current practice. The model has the advantage of analyzing alternate policies with respect to truck sizes. For example, reducing the ninety-five percent to ninety percent will, in effect, tend to allocate larger vehicles, and vice versa.

Indices	Distance Range (D) (Km)	Truck Size (T) (Tons)	Road Type(R)	Terrain Type(L)
1	0-15	Over 15	Limited Access Highway	Flat
2	15-50	9-15	Paved-High Standard	Hilly
3	50-200	5-9	Paved-Low Standard	Mountainous
4	+200	2-5	Unpaved	Urban

Note: Values for D, T, R and L are integers from 1 to 4 as specified above.

Table 3.5: Distance Traveled by Various-Sized Trucks on Different Road and Terrain Types

Backhaul Submodel

A backhaul submodel is essential to modeling freight movements in Korea because backhaul directly affects truck operating efficiency and the traffic to be carried by the road network. For a given pair of regions, the probability of backhauls is assumed to be a function of truck size and distance. The structure of the function and the calibration results for the coefficients are shown below.

The probability of backhauls is assumed to be a function of truck size and distance:

$$P^T = 1 - \exp(-K^T D) \tag{3.10}$$

where

P^T = backhaul probability of truck size T

K^T = coefficient for truck size T

D = fronthaul distance.

The model represented by equation (3.10) states that for a given size truck, the backhaul ratio will increase as the distance between the O/D pair increases. Coefficients, K^T, are calibrated as follows:

Truck Size (tons)	K^T
2-5	0.0009589
5-9	0.0019928
9-15	0.0024815
+15	0.0028522

Rail Empty Car Submodel

The problem of modeling rail empty cars is analogous to that of truck backhauls in nature. However, the greater simplicity of the rail network and the fact that the railroads have a single operator, the Korean National Railroads, imply that empty rail cars do not necessarily have to return to their starting points; if they do, they have to return by the "fronthaul" route. The only requirement is that, at any point in the rail network, the total number of cars coming in, either empty or full, must be equal to the total number going out for a given period of time. If this requirement is not met, surpluses or deficits of cars will accumulate. Thus, the objective of this submodel is to find movements of empty rail cars that minimize the cost of empty car movements subject to this requirement. The structure of the rail empty car submodel is

$$\min \quad \sum_{i \neq j} \sum_{j \neq i} c^{ij} E^{ij} \tag{3.11}$$

$$\text{s.t.} \quad \sum_{i \neq j} [F^{ij} + E^{ij}] = \sum_{j \neq i} [F^{ji} + E^{ji}] \tag{3.12}$$

$$E^{ij} \geq 0 \tag{3.13}$$

where

E^{ij} = number of empty rail cars to be hauled from node i to node j
F^{ij} = number of loaded rail cars to be hauled from node i to node j
c^{ij} = cost of hauling unit empty cars between nodes i and j.

3.3 Concluding Remarks

RDRI/KIST implemented the model introduced in this chapter and recommended a set of national transportation improvement programs to Korea's Transportation Subcommittee of the Fifth Five-Year Development Plan Committee in 1980. The following five criteria are used for evaluating alternative network systems generated by the Trip Generation and Network Design Model: (1) user benefits, (2) accessibility, (3) energy requirement, (4) vehicle utilization, and (5) cost and revenue. A diagrammatic presentation of the linkages among models described above is shown in Figure 3.2.

The chapter that follows describes processes for the identification, revision, and implementation of the national transportation improvement programs in Korea during the fifth FYDP period (1982–1986).

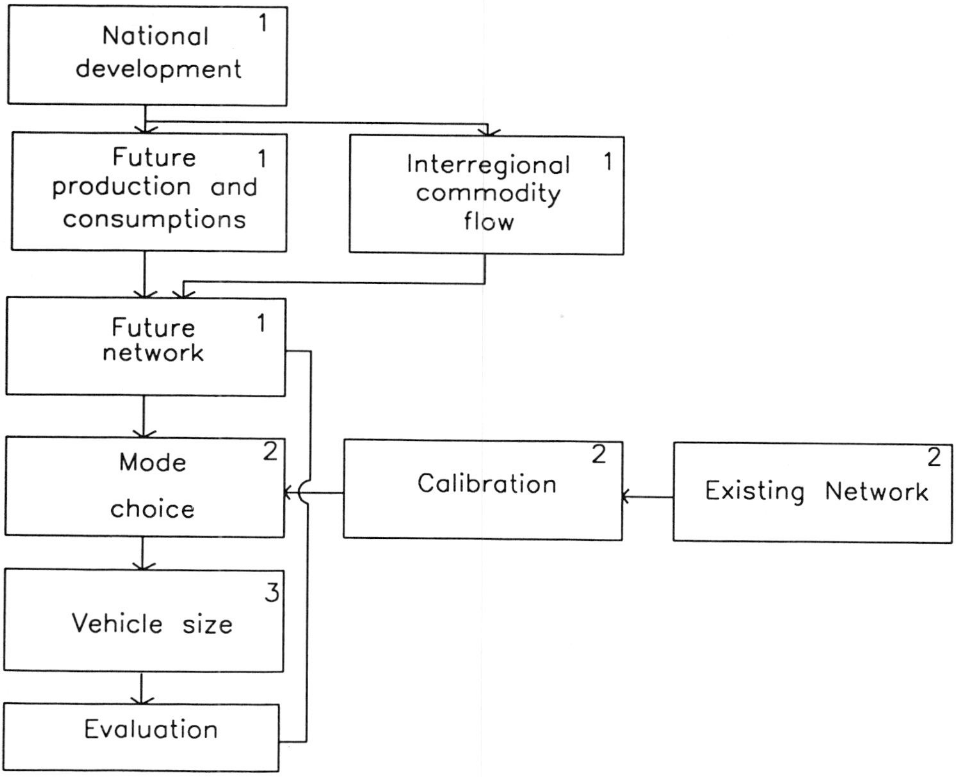

Figure 3.2: Linkages among RDRI/KIST Submodels

Chapter 4

Implementation of the Linear Programming Model: Experience from a Korean Application

4.1 Implementation of a Linear Programming Model:Issues and Implications for Model Building

Based on the results obtained from the model outlined in the previous chapter, RDRI/KIST recommended a set of national transportation improvement programs in 1980.[1] The model was solved using IBM's MPX360 package program. Detailed descriptions on recommended plans and programs are described in Kim(1980)[139]. Recommended additional networks to be completed by 1991 are shown in Figure 4.1.

In 1981, the government of Korea announced its original fifth FYDP. The notable difference between RDRI/KIST recommendations and the original fifth FYDP in the transportation sector was the addition of the new two-lane highway between Gwangju and Daegu in the fifth FYDP. After the decision was made that the 1988 Summer Olympic Games would be held in Seoul, additional four-lane highways were planned to connect the southern suburbs of Seoul to the northern part of Seoul.

[1] Portions of this chapter was written with Jong Gie Kim, and was published in *The Annals of Regional Science*, March 1985, 19:18–36.

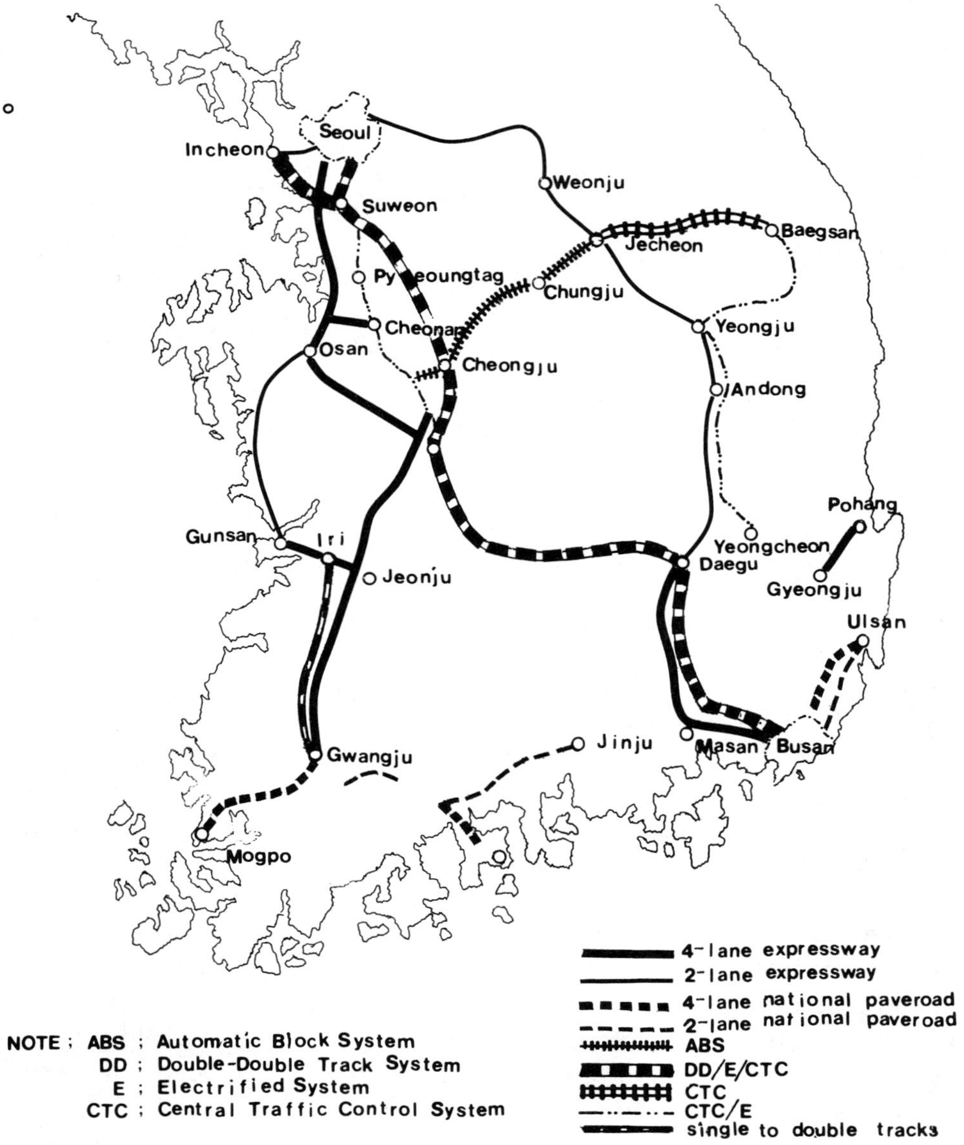

Figure 4.1: National Transportation Network Proposed by RDRI/KIST for 1991; Additions to Existing (1980) Network

Within three years after the improvement programs were recommended and one year after the adoption of the fifth FYDP, revisions of the programs became necessary due to changes in domestic and international economic variables. In 1983, the Economic Planning Board of Korea announced the revised plans in response to changes in the international and domestic economic climate. Three important international factors are cited as reasons: current decline in raw material prices, especially that of crude oil; the unforeseen delay in the world economic recovery; and an improvement in the national balance of payments (Suh 1983)[240].

Most of the revisions were made regarding the timing of investments. For example, extension of a double-track rail system from Iri to Gunsan was to be completed by 1988 instead of by 1985. The narrow-gauge rail system between Incheon and Suweon was modified to the standard-gauge system between 1985 and 1987 instead of the originally-planned period of 1983 to 1985. Similar revisions have been made for railroad improvements between Jecheon and Yeongju, Suweon and Daegu, and Jecheon and Baegsan.

The noteworthy difference between the original and revised programs is the emphasis on rural road improvements in the latter. By 1986 the total length of paved rural roads were to be doubled from 7.6 percent in 1982 to 13.5 percent, the percentage representing the length of paved roads out of total roads categorized as rural roads. Pavement ratio of intercounty roads (*Jibang Do*, in Korean) will be increased from 17.3 percent in 1983 to 29.2 percent in 1986. Pavement ratio of interprovincial road (*Guk Do*, in Korean) is planned to be increased from 67.5 percent in 1983 to 77.7 percent in 1986.

4.2 External Factors Affecting Program Formulation and Revisions

The decision to hold the 1988 Summer Olympics in Seoul was made in 1982 and has affected the programs for the transportation network development around the Seoul metropolitan area as well as outside. One result is the construction of "1988 Olympic Highway" between Daegu and Gwangju, connecting the southeastern region of the country to the southwestern region. The decision on the 1988 Olympic Highway can be considered one of the government's efforts to relieve pressure from the Gwangju region, which traditionally lags economically. In fact, the construction of the 1988 Olympic Highway played only a small role, if any, in the 1988 Summer Olympic Games because it did not affect accessibility from either Gwanju or Daegu to Seoul.

Other factors that caused the original fifth FYDP to be revised are three international factors, mentioned earlier, and some domestic changes in the economic climate between 1982 and 1983. As for the latter, the agricultural sector became a major domestic problem. Production increases in the agricultural sector have lagged far behind those of the manufacturing sector. While declining crude oil prices helped improve the nation's balance of payment, rice and fertilizer subsidies helped create larger budget deficits (Suh 1983)[240]. A possible alternative for solving this issue is to raise the opportunity for farmers' real income through the expansion of agriculture-related infrastructure (e.g., improving rural roads).

This suggestion is challenged by some analysts on the grounds that the rate of return from the public investment for improving urban roads is still greater than that from the investment for rural roads. Nevertheless, the government's efforts in raising the real income of farming sectors are well represented in the revised fifth FYDP.

Complex political factors are involved in the decision-making processes for formulating national transportation improvement programs. Though the complicated politics are beyond what the model builders can represent, the presence of these factors becomes inevitable in practice. Model builders therefore usually have very little to say about political decisions and accept them as they are handed down.

The effects of rapid changes in international and domestic economic climates upon Korean national transportation improvement programs suggests that a model for national transportation development planning should be made so that those external changes can be flexibly brought into it. This requirement becomes important in building a model for developing countries, whose economic climates are becoming more and more dependent on changes in international economic conditions. More specifically, the macroeconomic model and network generation model have to be explicitly expressed in order to represent the effect of external economic changes on the network system.

Regarding this, Friesz, Tobin, and Harker (1983)[104] advocated the simultaneous treatment of macroeconomic and network generation models and have shown in earlier studies the possibility of building a simultaneous system of macroeconomic and network generation models through a generalized spatial price equilibrium approach (Friesz, Gottried, and Morlok 1981[101]; Tobin and Friesz 1983[103]; Harker 1983[114]). Kim, Boyce, and Hewings (1983)[138], on the other hand, have handled simultaneous treatment of macroeconomic and network generation models through a network equilibrium approach.

Both frameworks of the spatial price and network equilibrium approaches, however, have common critical problems: (1) alternative networks have to be given a priori, and (2) assumptions associated with the capacity of each link for each mode become extremely sensitive to the model results. The first problem poses two further questions: (1) How should these alternative networks be generated? and (2) How many alternatives should be tested before one can reasonably conclude that a specific network is the best one?

The first question is concerned with modeling network design. The network design problem has been formulated either in terms of discrete links to be considered for addition to an existing network, or in terms of increases in each link's capacity in a continuous manner (Beckmann 1967 [23]; Boyce and Soberanes 1979 [43]; Marble and Garrison 1958 [171]; Quandt 1962 [207]; Scott 1969 [221]; Stair 1967 [231]; Werner 1968 [257]; Wong 1983 [262]). Obviously, both formulations require rather severe abstractions from problems faced by transportation planners. A critical shortcoming in existing models for network design is the absence or the lack of feedback interactions between transportation activities and the rest of economic activities. If an alternative network is derived without much attention to macroeconomic variables and yet is used as an input for a simultaneous analysis of a macroeconomic and network performance, a vicious cycle will most probably occur.

The second question concerns the validity of an "optimum" when the equilibrium volume is obtained for a given network. There is no way of knowing whether the "optimum" obtained is actually a local or global optimum, since an infinite number of alternative networks can be generated for a nation.

The second problem has to do with defining and estimating the the capacity of each link. Unlike urban networks, where well-defined peak hours exist, no common delivery time can be found in interregional freight shipment patterns. This leads to a difficulty in defining capacity of a link. The capacity, usually defined by volume and day, would vary widely depending on hours per day of service. Consequently, the average link cost and delay functions, such as those derived in CACI (1980)[53], also vary widely depending on the definition of capacity of a link. For example, a link may be interpreted to be congested if shipment is assumed to be allowed only for eight hours per day, but may be uncongested if shipment is assumed to be allowed all day. The alternative definition of capacity affects estimated shipment costs, which, in turn, affect the results on mode choice, route choice, and, ultimately, the contents of national transportation development plans.

Models for interregional commodity flows such as the one developed by

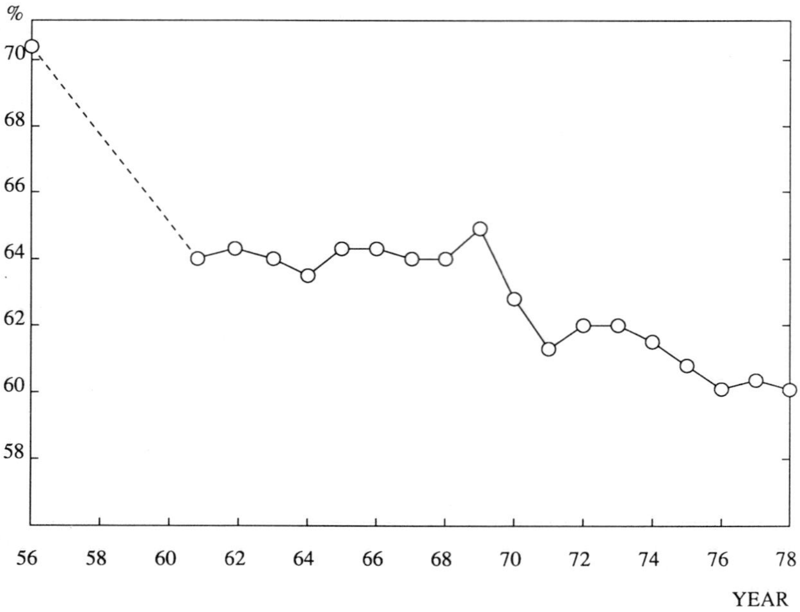

NOTE : The Percentage of Loaded Rail Cars Indicates the Rate of Utilization of Loaded Cars Among Total Rail Cars Being Used for a Given Year.

Figure 4.2: Percentage of Loaded Rail Cars against Total Rail Cars in Korea (1956–1978)

RDRI/KIST also have many practical problems in them. Except in the simplest situations, data on transportation costs from region i to region j does not exist. What does exist are transportation rates for individual carriers for specific routings. What rates are actually used in commodity shipments only become known in the course of decisions regarding actual mode and route choice by individual shippers.

The importance of developing and including a backhaul model for national transportation planning in Korea becomes clear historic trends of empty rail car movement (Figure 4.2). The figure seems to show, at first glance, that the Korean National Railroad's operation is becoming less efficient. Close examination of the Korean National Railroad's operation, however, indicated that the value of the time necessary for shipping commodities has been increasing. In the 1950s, the empty car would have waited until it was loaded. Now, commodities tend to be shipped as soon as possible, even at the cost of increasing the number of empty cars transferred to those points where the shipping demand exists.

Generalization of this phenomenon implies that for most developing countries, particularly those countries where regional imbalance of growth is imminent (e.g., the Seoul and Busan growth poles in Korea), a backhaul model becomes crucial for representing movement of vehicles and fleets among regions. Regarding the vehicle size model, we should know that in developing countries small trucks are still dominant in freight shipment on roads, because drivers' wages are usually not an important variable in composing transportation policy models. Wages affect operating costs, energy consumption, maintaining cost of networks, and, ultimately, planning and designing network systems.

4.3 Concluding Remarks

Building a multiregional multicommodity model for the national transportation improvement programs is complex. A transportation model that is to be useful for the formulation of national development plans must adequately show the mutual interaction between improvement of the transportation network and economic activities of private sectors. Showing this interaction is important because the location and levels of production and consumption, as well as origins and destinations of passenger trips, are affected by the provision of transportation facilities. At the same time, the location and levels of private activities affect the demand placed on the transportation system. It matters a great deal how much and where transportation is provided in the multiregional economic system. This point becomes more critical in developing countries, where resources are limited, changes in external forces are ever-present and the existing transportation planning system is often inadequate.

From this viewpoint, an appropriate model for national transportation planning in developing countries is a network design model that has explicit linkages between international and interregional economic variables. Such models are introduced in the chapters that follow.

Part III

NONLINEAR PROGRAMMING MODELS AND APPLICATIONS

Chapter 5

Combined Input-Output and Commodity Flow Model: A Nonlinear Programming Formulation

5.1 Introduction

Many developing nations have formulated and implemented national development programs to increase their national well-being.[1] Korea began its first five-year development program in 1961 and successfully implemented its fourth FYDP in 1981, as described in previous chapters.

The planning process for formulating FYDPs is complex. The process often begins with the establishment of goals: the main goal is usually set in terms of gross national product (GNP), which in turn is translated into the amount of goods to be produced, exported, and imported in each sector.

In planning and implementing development programs, planners' attention has been attracted to, among other issues, the establishment of a proper relationship between the transportation sector and all other sectors. Planners are concerned not only with the proper distribution of scarce resources among sectors, but also with the trade-offs between development costs of certain regions and transportation costs that provide accessibility to them.

[1]This chapter, in its original form, was written with David E. Boyce and Geoffrey J.D. Hewings, and was published in *Geographical Analysys*, 1983 15(4):330–342.

This issue is known in urban areas as transportation-land use interactions.

In the last two decades, a number of approaches have been developed, ranging from single-region economic base models to large-scale multiregional input-output, linear programming, and econometric models. Most of the existing models, however, require that each subregion's production or consumption, or both, are known. This assumption presents problems for policy decision makers who are implementing proposed development programs. Particularly in developing countries, where resources are sometimes scarce, decision makers want to know the proper amount of production to be allocated to each subregion; they want to enhance the well-being of the nation or the entire system of regions. Moreover, there is no way of ascertaining the proper distribution of scarce resources among sectors, nor of weighing trade-offs between the costs of developing certain regions and the transportation improvement costs of providing accessibility to them.

The open nature of regional economies, both in developed and developing countries, has presented an additional difficulty for the design and implementation of multiregional analytical models. While the core of development programs in developing countries relies heavily on the development of the export sector, most existing models cannot suggest the trade-offs among port development, regional development that produces export goods, and transportation development that connects the two.

The purpose of this chapter is to propose for an open economic system a multiregional, multicommodity model that can be useful in determining optimum locations of economic activities and the transportation improvements necessary to serve them so as to maximize the well-being of the entire system. The chapter provides some possible directions for the empirical implementation of an entropy-constrained formulation within the context of a multiregional commodity flow/input-output model proposed by Wilson(1970)[259].

5.2 Interregional Input-Output Models

Interest in interregional commodity flow and input-output modeling has been a consistent part of the general literature on regional and interregional analysis. The early work of Isard (1951)[126] and Moses (1955)[186] has been extended through the use of more sophisticated matrix algebra techniques. These models may be grouped into three categories: (1) allocation/RAS types (Stone et al. 1963[234]; Bacharach 1970[11]; Nevin, Roe, and Round

1966[194]; Round 1972[217]; Corbis and Vallet 1976[68]); (2) linear programming models (Hewings 1970[119];Mathur 1972[176]; and the one introduced in Chapter 3); and (3) gravity-type models such as the Leontief-Strout (1963)[164] model. This last type has been modified, expanded, and applied to a large number of problems by Polenske (1970a[201], 1970b[202]). Unlike the Nevin, Roe, and Round (1966)[194] model, in which trade in each commodity can only occur in one direction, Leontief and Strout view regional supply and demand as being comprised of a set of goods whose origins may vary. Producers and consumers are assumed to be indifferent about these origins and, as a result, the model envisages shipments taking place within one region's supply pool. No attempt is made to distinguish between local supplies and supplies from other regions. Regional outflows of goods vary directly with the level of output in the producing region, while regional inflows vary indirectly with the level of output in the producing region. At the same time, regional inflows vary directly with regard to regional consumption. Interregional flows are, however, inhibited by the "friction of distance" effect, which means that interaction varies inversely with distance.

Leontief and Strout performed several tests of their model; however, the greatest use of this model, including a number of derivatives, was made by Polenske (1972)[203]. Her multiregional input-output model was first implemented in 1963 and subsequently updated in 1967. The models have been used for transportation impact analysis, energy demand studies, forecasting of occupational needs, and, most recently, in conjunction with a demographic model for forecasting activity levels for a small set of states (Evans and Baxter 1980)[89].

The general approaches described above consider interregional trade from an input-output perspective, namely, the disaggregation of technical coefficients into a set of local and nonlocal components. The models may be regarded as models of interregional trade rather than of interregional commodity flow. The latter models may be thought of as derivatives of transportation models, where the major focus of interest is on the specification of a transportation network over which the commodities may flow and on the derivation of the choice of mode to be used in this network. In these models, supply and demand equations are not usually associated with interindustry models, although the degree of linkage is considerable and of great interest. Transportation system models are important as a means of evaluating the impacts of proposed policy changes on the flows of commodities between regions.

5.3 Commodity Flow Models

That interregional commodity flows are directly proportional to regional production and consumption and inversely proportional to transportation cost (or its surrogate, distance) has been observed by many researchers over several decades. These relationships have often been criticized as exhibiting only empirical regularities in the data and not providing any rigorous theory of spatial interaction. A major step toward such a rigorous derivation of these gravity relationships from more basic precepts was provided by Wilson's application of entropy-maximizing methods to commodity flow analysis (1970)[259]. Subsequently, Erlander (1977)[85] showed how the same problem could be given a macrobehavioral interpretation without appealing to the concepts of information theory.

Wilson also showed how input-output models of a regional economy could be combined with the gravity model to provide the most general model of the interregional economy proposed so far. He derived a family of four commodity flow models, each of which differed only in terms of cost minimization (e.g., see Harris 1974)[116]. If regional production and consumption are given, then this results in the transportation problem of linear programming results as follows.

Let

c_r^{ij} = transportation cost from region i to region j for commodity r

X_r^{ij} = amount of commodity r shipped from region i to j

C_r = total cost of transporting commodity r in the system

X_r^i = production of commodity r in region i

y_r^i = consumption of commodity r in region i.

An efficient allocation of production to consumption may be found by

$$P5.1) \quad \min \quad C_r = \sum_i \sum_j c_r^{ij} x_r^{ij} \tag{5.1}$$

$$\text{s.t.} \quad \sum_j X_r^{ij} = X_r^i \text{ for all } i \tag{5.2}$$

$$\sum_i X_r^{ij} = Y_r^j \text{ for all } j \tag{5.3}$$

$$X_r^{ij} \geq 0 \text{ for all } i \text{ and } j. \tag{5.4}$$

Solution of this problem results in the identification of, at most, $2n - 1$ positive flows, where n is the number of regions. The cost-minimizing solution does not allow for shipments between two regions in both directions, since, in principle, one of these shipments would be unnecessary or inefficient.

Observation of commodity flows in the real world clearly indicates that the cost-minimizing solution does not describe the actual situation. More than $2n - 1$ flows occur, and often commodities in the same class move in both directions between two regions. This latter phenomenon, known as crosshauling, is a result of two conditions. First, the level of aggregation in such analyses is often high, so that opposing shipments of distinctly different goods in the same commodity class may appear as crosshauling. Second, actual crosshauling of the same commodity may occur because of established trading patterns, lack of information, product differentiation (through advertising), and other institutional factors.

Because of the dilemma presented by crosshauling, some method is needed that will modify the transportation problem formulation to provide a more realistic description of the commodity flow pattern. Erlander (1977)[85] proposed that the entropy function

$$S_r = -\sum_i \sum_j X_r^{ij} \ln(X_r^{ij}) \tag{5.5}$$

can be used to accomplish this objective. S_r can be interpreted as a scalar measure of the level of dispersion, or crosshauling, in an observed pattern of crosshauling in both of the senses discussed above. By adding constraint (5.5) to the above transportation problem, the solution is constrained to produce a flow matrix that has the same level of crosshauling as the observed matrix. The model would now become as follows:

$$\begin{array}{ll} P5.2) \quad \min & (5.1) \\ \text{subject to} & (5.2) - (5.5). \end{array}$$

The solution of this constrained optimization problem may be shown to yield a doubly-constrained gravity model of the same form as that derived by Wilson (1970)[259] and Wilson et al. (1981)[261].

The possibility of interpreting the Lowry model (1964)[167] as an input-output framework was implied in the works of Broadbent (1973)[47] and Romanoff (1974)[216]. The actual modification, however, was developed by Macgill(1977)[168]. As in the original Lowry model, Macgill assumed that there exists an a priori determined number of "basic" employees in each zone.

The activity relationships between the basic and retail-service (nonbasic) sectors were traced by the household sales of labor to both sectors and retail sales to household. These variables become the bases for formulating Leontief-type demand-driven input-output coefficients. The author further extended a spatially desegregated input-output model as well as a maximum-entropy model in a Lowry-type framework.

5.4 Network Design Models: State-of-the-Art

Models for formulating investment priorities for road improvements have not received much attention, nor have models of the network design problem that could shed light on investment priority issues. As mentioned before, the network design problem has been formulated either in terms of discrete links to be considered for addition to an existing network or in terms of increases in each link's capacity, defined as continuous (e.g., Marble and Garrison1958[171]; Quandt 1962[207]; Beckmann 1967[23]; Werner 1968[257]; Stair 1967[231]; Ridley 1968[211]; Scott 1969[221]; Boyce and Soberanes 1979[43]; Wong 1983[262]). Both formulations represent rather severe abstractions from the problems faced by transportation planners.

A critical shortcoming of models of transportation network investment problems (e.g., Bergendahl 1969[30]; Barber 1975[14], 1977[15]; Ahmed et al. 1976[4]) and models of network design in general is the lack of representation of interaction between transportation and other economic sectors. Most existing models miss the essential normative issue, namely the interaction between public and private sectors. Recent works by Ralston and Barber (1982[209], 1984[208]) present an important theoretical framework that allows for a large-scale network, an area of research that has yet to be undertaken.

The assumption that interregional commodity flows are proportional to regional production and consumption and inversely proportional to transportation costs was often criticized as merely reflecting empirical regularities in the data; that is, it was not derived from any rigorous theory of spatial interaction. A major step toward such a rigorous derivation of these "gravity" relationships from more basic precepts was provided by Wilson's (1970)[259] application of entropy maximizing methods to commodity flow analysis. Wilson further demonstrates how input-output models of a regional economy could be combined with the gravity model to provide a general model of the interregional economy. Although the original model itself is not a network equilibrium model, it has had a significant impact upon subsequent models in this category.

The problem of network equilibrium for a single mode was first formulated *mathematically* by Beckmann, McGuire, and Winsten (1956)[24]. An efficient computational algorithm was devised by LeBlanc, Morlok, and Pierskalla (1975)[160]. The equilibrium is characterized by Wardrop's (1952)[252] two principles:

1. The shipments costs over all paths that are used from each origin to each destination are equal.

2. No unused path has a lower shipment cost.

Thus, at equilibrium, no shipper can lower costs by switching paths.

In the model developed by Roberts (1966)[213] and Kresge and Roberts (1971)[151], later known as the Harvard-Brookings model, each shipper chooses the shortest path from an origin to a destination, the amount shipped between an O/D pair being determined by a Koopmans-Hitchcock submodel. Peterson and Fullerton (1975)[200] presented an optimization model that incorporated nonlinear congestion cost functions searching for Wardrop's equilibrium conditions. Devarajan (1981)[80] discusses how user-equilibrium (Wardrop's first principle) is a form of the Nash (1951)[193] equilibrium in a noncooperative game. Friesz, Gottfried, and Morlok (1981)[101] present a model in which two distinct groups of agents (shippers and carriers) are first recognized in modeling a network-specific commodity flow problem. The model assumes that shippers act according to Wardrop's first principle of user optimization, while carriers behave according to Wardrop's second principle of system optimization. They modeled these two distinctive behaviors sequentially, first solving the shippers' submodel, then loading the results on the carriers' submodel. As Harker (1983)[114] and Charles River Associates (Egan 1982)[83] noted, however, a sequential approach does not guarantee consistency between the two submodels.

The combined decision of a shipper and carrier is represented in the models developed by CACI (CACI 1980[53]; Bronzini 1980[48], 1982[49]); USDOT (Swerdloff 1971[242]); Green (1980)[111]; Kornhauser (1982)[150]; and Kim, Boyce, and Hewings (1983)[138]. Kornhauser has developed an interactive model of the freight system in which network cost parameters can be altered in such a way that predicted flows are close to replicating observed data on the network. The models have proven to be useful tools for decision makers. Nearly all models in the network equilibrium category ignore, however, the question of the source of transportation cost data. Except in the simplest situation, data on transportation costs between regions do not exist. What do exist are transportation rates for individual modes for specific routings. The rates actually used in commodity shipments only become known in the course of decisions regarding actual mode and route choice (Boyce and Hewings 1980)[39].

5.5 Combined Regional Input-Output and Interregional Commodity Flow Model for Unknown Regional Production

The model reviewed above can be combined with a system of regional input-output models by adding the basic input-output relationship between interregional commodity production and consumption. To form a combined model when total production of commodity r at region i is unknown, two basic assumptions are necessary: (1) total exports of commodity $r(E_r)$ and imports (I_r) for the system of regions are known, and (2) only transportation costs determine interregional flows. In this mode, final demand (y_r^i) is also assumed to be known.

The model has already been introduced in chapter 3, where total transportation costs are minimized subject to input-output and export-import constraints. The basic structure of the model may be stated as follows:

$$P5.3) \quad \min C = \sum_r \sum_i \sum_j c_r^{ij} X_r^{ij} + \sum_r \sum_{i \in p} d_r^i (E_r^i + I_r^i) \tag{5.6}$$

$$\text{s.t.} \sum_{i \in p} E_r^i \geq E_r \text{ for all } r \tag{5.7}$$

$$\sum_{i \in p} I_r^i \geq I_r \text{ for all } r \tag{5.8}$$

$$\sum_j X_r^{ij} + I_r^i + X_r^i = \sum_j X_r^{ij} + \sum_q a_{rq}^i \left(\sum_j X_q^{ij}\right) + y_r^i + E_r^i \text{ for all } r, i \tag{5.9}$$

$$-\sum_i \sum_j X_r^{ij} \ln(X_r^{ij}) \geq S_r \text{ for all } r \tag{5.10}$$

$$X_r^{ij} \geq 0 \text{ for all } r, i, \text{ and } j$$

$$E_r^i \geq 0 \text{ for all } r \text{ and } i$$

$$I_r^i \geq 0 \text{ for all } r \text{ and } i$$
$$X_r^i \geq 0 \text{ for all } r \text{ and } i. \qquad (5.11)$$

E_r^i and I_r^i are the amounts of exports and imports of commodity r through ports in region i that are to be estimated. The symbol p denotes the subset of regions in the system of regions that have port facilities. The cost of exporting or importing one unit of commodity $r(d_r^i)$ is assumed to be known and includes the cost of shipping through ports in region i plus any differential cost above or below a base shipping cost.

Objective function (5.6) represents total interregional transportation and export/import costs. Equations (5.7) and (5.8) ensure that export targets are met and import quotas are not exceeded.

The LHS of equation (5.9) consists of flows of commodity r into region r, including flows from region i itself, imports of r into region i, and total production of r in i. Equation (5.9) requires that these inflows of r into i equal the flows of r into other regions, the intermediate and final demand for r in i, and exports of r from i. The intermediate demand for commodity r is represented by summing the requirements for commodity r in producing one unit of commodity q, a_{rq}^i, multiplied by the total production of commodity q in region i, $\sum_j x_q^{ij}$. The final demand for commodity r in region i is y_r^i, and this is the only required known variable in equation (5.9) except for the input-output coefficients.

Constraint (5.10) represents the level of crosshauling of commodity r in the system of regions. Constraint (5.11) requires that the unknown variables be nonnegative.

We seek to minimize (5.6) subject to (5.7)-(5.11). To this end, we form the Lagrangian L as follows:

$$\begin{aligned} L = C &+ \sum_r \sigma_r(E_r - \sum_{i\in p} E_r^i) + \sum_r \rho_r(-I_r + \sum_{i \in p} I_r^i) \\ &+ \sum_r \sum_i \gamma_r^i (\sum_q a_{rq}^i (\sum_j X_q^{ij}) + y_r^i + E_r^i - \sum_j X_r^{ij} - I_r^i) \\ &+ \sum_r \frac{1}{\mu_r}(S_r + \sum_i \sum_j X_r^{ij} \ln(X_r^{ij})) + \sum_r \sum_i \sum_j \lambda_r^{ij}(-X_r^{ij}) \\ &+ \sum_r \sum_i \lambda_{r1}^i(-E_r^i) + \sum_r \sum_i \lambda_{r2}^i(-I_r^i), \end{aligned}$$

where $\sigma_r, \rho_r, r_r^i, 1/\mu_r, \lambda_r^{ij}, \lambda_{r1}^i$, and λ_{r2}^i are Lagrange multipliers associated with constraints (5.7), (5.8), (5.9), (5.10), and (5.11). By differentiating

L with respect to the unknown variables X_r^{ij}, E_r^i, and I_r^i, we obtain the following optimality conditions:

$$\begin{aligned}
\frac{\partial L}{\partial X_r^{ij}} &= c_r^{ij} + \sum_{q=1}^{n} \gamma_q^i a_{rq} - \gamma_r^j + \frac{1}{\mu_r} \ln(X_r^{ij}) - \lambda_r^{ij} = 0 \\
\frac{\partial L}{\partial E_r^i} &= d_r^i - \sigma_r + \gamma_r^i - \lambda_{r1}^i = 0 \\
\frac{\partial L}{\partial I_r^i} &= d_r^i - \rho_r - \gamma_r^i - \lambda_{r2}^i = 0 \\
\text{and } \lambda_r^{ij} X_r^{ij} &= 0 \qquad \sum_r \sigma_r(-\sum_{i \in p} E_r^i + E_r) = 0 \\
\lambda_{r1}^i E_r^i &= 0 \qquad \sum_r \rho_r(\sum_{i \in p} I_r^i - I_r) = 0 \\
\lambda_{r2}^i I_r^i &= 0 \sum_r \frac{1}{\mu_r}(S_r + \sum_i \sum_j X_r^{ij} \ln(X_r^{ij})) = 0.
\end{aligned}$$

These conditions may be interpreted as follows:

1. If $X_r^{ij} > 0$, then $\lambda_r^{ij} = 0$ and $\ln X_r^{ij} = \mu_r(\gamma_r^j - \sum_q \gamma_q^i a_{rq}^i - c_r^{ij})$.

   Moreover, assuming c_r^{ij} is finite, then $\ln X_r^{ij} > -\infty$,
   so X_r^{ij} is never zero.
2. If $E_r^i > 0$, then $\lambda_{r1}^i = 0$ and $\gamma_r^i = \sigma_r - d_r^i$.
   If $E_r^i = 0$, then $\lambda_{r1}^i \geq 0$ and $\gamma_r^i \geq \sigma_r - d_r^i$.
3. Likewise, if $I_r^i > 0$, then $\lambda_{r2}^i = 0$ and $\gamma_r^i = \rho_r + d_r^i$.
   If $I_r^i = 0$, then $\lambda_{r2}^i \geq 0$ and $\gamma_r^i \leq \rho_r + d_r^i$.

Further economic implications can be stated from statements 2 and 3 above by interpreting Lagrangian multipliers associated with export- and import-related equations, as follows (see Coelho and Williams 1978[65] for interpreting locational surplus in a similar context):

σ_r: the opportunity cost of exporting an additional unit of r from the entire system of regions ($= \partial L^*/\partial E_r$)

ρ_r: the opportunity cost of importing an additional unit of r to the entire system of regions ($= \partial L^*/\partial I_r$)

γ_r^i: the locational surpluses arising from producing r at export region i

d_r^i: export (import) cost of unit r at i export-import region.

Statement 2 further implies that the locational surpluses arising from producing r at $i(\gamma_r^i)$ should at least equal the difference between the opportunity cost of exporting $r(\sigma_r)$ and the unit export cost (d_r^i), if commodity r is to be exported through i. If the location surpluses at export region i are larger, that is , $\gamma_r^i > \sigma_r - d_r^i$, the complementary slackness theorem implies that exports should occur elsewhere, since $(\partial L/\partial E_r^i)E_r^i = 0$ at optimum.

Likewise, statement 3 further implies that the locational surpluses arising from producing r at import region $i(\gamma_r^i)$ should be at most the same as the sum of the opportunity cost of importing additional r to the entire system of regions (ρ_r) and the unit import cost of r at d_r^i, if commodity r is to be imported through i. Otherwise, imports will take place elsewhere.

Returning to the first condition and solving for X_r^{ij} (see Appendix A),

$$X_r^{ij} = \exp\,(\mu_r \gamma_r^j - \mu_r \sum_q \gamma_q^i a_{rq}^i - \mu_r c_r^{ij}).$$

Letting

$$E_r^j = \exp(\mu_r \gamma_r^j)$$

and

$$\sigma_r^i = \exp\,(\mu_r \sum_q \gamma_q^i a_{rq}^i),$$

we obtain

$$X_r^{ij} = \sigma_r^i E_r^j \exp(-\mu_r c_r^{ij}), \tag{5.12}$$

which is an unconstrained spatial interaction model with attractiveness measures defined by material balance equation (5.9).

The import and export variables may be determined once (X_r^{ij}) is known. Equation (5.12) is an entropy-constrained version of a spatial interaction model that accounts for interindustry flows. The deterrence parameter, μ_r, relating commodity flow to transportation cost, may be determined in relation to the assumed level of crosshauling, S_r. Equation (5.12) is analogous to the Leontief-Strout formulation. The above model can be extended to include production costs and costs of increasing production capacity. Port capacity constraints could also be included.

5.6 Extensions to Include Network Representations of Transportation Cost

Nearly all models of commodity flow, including Wilson's, ignore the question of the source of transportation costs required as one of the major inputs of the model. Except in the simplest situations, data on transportation costs from region i to j do not exist. Still, transportation rates for individual modes (e.g., rail, motor carrier, pipeline, barge) for specific routings do exist. Unfortunately, even these rates are not readily accessible to the modeler or analyst in most situations. The rates actually used in commodity shipments become known only in the course of decisions regarding actual mode and route choice.

A useful point of departure for examining route and mode choice models is to consider route choice for a transportation system comprised of one mode, like rail. In the absence of origin-destination rate information for alternative routes, and abstraction from the issues of the origination carrier's control over route choice (Lansdowne 1979)[155], assume that each link of the network has a unit user cost c_a, which is an increasing function of the annual flow f_a:

$$c_a = c_a(f_a).$$

The behavior of the shipper in this situation may be generally depicted in terms of cost minimization. If link costs are fixed, the shipper will choose the path or sequence of links through the network that has the minimum total cost. If congested conditions exist in the network, then the transportation cost on each link is not fixed, but may be regarded as an increasing function of the flow on that link. Since any given link may serve many origin-destination pairs, the problem is highly interdependent. If, in this situation, shippers continue to seek their minimum cost paths, then a network equilibrium problem results. The mathematical formulation of the problem follows from the result that at network equilibrium the total area under the link cost functions is minimized. The network equilibrium and the versions of Wilson's model introduced above can be combined as follows:

$$P5.4) \quad \min \sum_a \int_0^{f_a} c_a(x)dx \; + \; \sum_r \sum_i d_r^i (E_r^i + I_r^i) \qquad (5.13)$$

$$\text{s.t. } X_r^{ij} \; = \; \sum_p X_r^{ijp} \text{ for all } i, j, r \qquad (5.14)$$

$$f_a = \sum_r \sum_i \sum_j \sum_p X_r^{ijp} \delta_{ar}^{ijp} \tag{5.15}$$

$$X_r^{ijp} \geq 0 \text{ for all } i, j, p, r \tag{5.16}$$

$$S_r \geq -\sum_i \sum_j X_r^{ij} \ln(X_r^{ij}) \text{ for all } r \tag{5.17}$$

$$\sum_{i \in p} E_r^i \geq E_r \text{ for all } r \tag{5.18}$$

$$\sum_{i \in p} I_r^i \leq I_r \text{ for all } r \tag{5.19}$$

$$\begin{aligned} \sum_j X_r^{ij} + I_r^i &+ X_r^i = \sum_j X_r^{ij} + \\ \sum_q a_{rq}^i (\sum_j X_r^{ij}) &+ y_r^i + E_r^i \text{ for all } r, i \end{aligned} \tag{5.20}$$

$$\begin{aligned} X_r^{ij} &\geq 0 \text{ for all } r, i, j \\ E_r^i &\geq 0 \text{ for all } r, i \\ I_r^i &\geq 0 \text{ for all } r, i \\ X_r^i &\geq 0 \text{ for all } r, i, \end{aligned} \tag{5.21}$$

where

X_r^{ijp} = amount of commodity r shipped from region i to region j on path p

f_a = total flow on link a

c_a = cost of link a

$$\delta_{ar}^{ijp} = \begin{cases} 1, & \text{if link } a \text{ is included in path } p \text{ from } i \text{ to } j \text{ for commodity } r \\ 0, & \text{otherwise.} \end{cases}$$

The solution to this problem is often referred to as a "user-equilibrium" solution. Unless link costs are constant, this solution does not minimize total transportation costs.

5.7 Limited Application of the Model to Korea

As an initial step toward applying the proposed model to Korea, the level of dispersion of commodity flows X_r^{ij} (equation (5.10)) was estimated using an observed flow matrix in 1985. The commodity flow matrix was compiled by the Korea Institute of Science and Technology, which collected interregional flows of eighteen commodities to and from 100 regions in Korea. (For the detailed map of the 100 regions, see Kim (1980)[139]).

The original eighteen commodities were grouped into four commodity groups based primarily on transportation modes used for shipping them in 1978, as follows:

Group 1:
1 Coal
2 Limestone
3 Petroleum
4 Metallic minerals

Group 2:
10 Cement
11 Nonmetallic minerals

Group 3:
14 Agricultural products
15 Livestock and silkworms
16 Forest products
17 Fish

Group 4:
5 Food and tobacco
6 Textile
7 Lumber, wood, and furniture
8 Pulp and paper
9 Chemicals
12 Machinery
13 Miscellaneous manufacturing products
18 Other

For computational convenience, equation (5.10) was normalized as follows:

$$S_r^{obs} = -\sum_i \sum_j \frac{X_r^{ij}}{\sum_i \sum_j X_r^{ij}} \ln\left(\frac{X_r^{ij}}{\sum_i \sum_j X_r^{ij}}\right) \quad (5.22)$$

$$X_r^{ij} = A_r^i X_r^i B_r^j Y_r^j \exp(-\beta_r c_r^{ij}), \quad (5.23)$$

where $X_r^i = \sum_j X_r^{ij}$, $Y_r^i = \sum_i X_r^{ij}$ and both are assumed to be known and S_r^{obs} is the observed level of crosshauling. The values of the transportation deterrence parameter in equation (5.23), β_r^{obs}, were estimated by iterative

Commodity Group (h)	S_h^{OBS}	β_h^{OBS}	Chi-Square	Regression Analysis ($X_h^{JL\cdot EST}$ against $X_h^{JL\cdot OBS}$)		
				a	b	R^2
1	5.30075	1.83300	$6,659.6 \times 10^6$	1,818.26367	0.65944 (85.13205)	0.68795
2	4.79244	1.94442	$1,136.4 \times 10^6$	1,460.66943	0.49912 (57.37292)	0.57488
3	5.57108	1.0553	$28,010.3 \times 10^6$	2,422.12391	0.17472 (16.39162)	0.25079
4	5.72683	2.02776	$159,048.8 \times 10^6$	5,730.50391	0.40102 (43.96158)	0.50361

Table 5.1: Calibration Results of the Commodity Flow Model for Korea

methods, as suggested by Williams (1977)[258]. Both values of S_r^{obs} and β_r^{obs} are shown in Table 5.1 by commodity group.

The relationships between β_r and X_r^{ij} are such that when $\beta_r = 0, X_r^{ij}$ is proportional to production and consumption and unrelated to transport costs. As β_r approaches infinity, X_r^{ij} tends to the solution of the original transportation problem. The observed values shown in Table 5.1 indicate that shippers of commodity group 1, mostly manufacturing commodities, are the most sensitive to transportation costs in Korea. Commodity groups 1 and 2, which are mostly minerals, have shippers that are also relatively sensitive to transportation cost. Shippers of commodity group 3, which is agricultural products that are available locally in most regions in Korea, are least sensitive to transportation cost.

Also shown in Table 5.1 are evaluative measurements of the model (χ^2) and regression analysis of the estimated commodity flow ($X_r^{ij\cdot est}$) against observed ($X_r^{ij\cdot obs}$). χ^2 values were calculated as follows:

$$\chi_r^2 = \sum_i \sum_j \frac{(X_r^{ij\cdot est} - X_r^{ij\cdot obs})^2}{X^{ij\cdot EST}}.$$

A linear regression analysis was performed as the second evaluative measurement for the performance of the model as follows:

$$X_r^{ij\cdot est} = a + bX_r^{ij\cdot obs}.$$

The values of a, b, and R^2 are shown in the last three columns of the table. The R^2s indicate that the model performs reasonably well for commodity groups 1, 2, and 4, while its low value for commodity group 3 may occur because most demands for agricultural products are locally supplied, and relatively less crosshauling is involved.

5.8 Concluding Remarks

One of the important questions faced by decision makers for national and regional development in most countries, particularly in developing countries, is the proper allocation of scarce capital and other resources between regional development and infrastructure improvements. The efficient allocation of resources among alternative transportation systems (for example, rail, highways, waterways, and terminals, including ports) is another important issue.

Most existing analytical models are limited (Barber 1975[14], 1977[15]). There is no way of efficiently allocating resources among sectors, since the amounts, location, and types of production must be known a priori. The assumption of constant transport cost among regions, regardless of volume to be shipped, is of little or no value to planners and decision makers when assigning priority for transportation improvements.

The solution of the proposed model provides a more complete description of the types, amounts, and location of production (X_r^i) and the amounts of efficient transportation systems improvements (X^{ijp}), which are based on either user-equilibrium or system-optimal solutions.

While the model introduced above provides a more complete description of the issues involved in modeling multiregional-multicommodity flow problems, particularly in a developing country's framework, a number of further extensions of the model can be made.

1. Exogenous treatment of the regional final consumption, y_r^i, can also be treated endogenously by constructing input-output models that include the final consumption sector. Such endogenous treatment of the final consumption can be found in the models of Mills (1975)[179] and Kim (1979)[134] in an urban context, and the social accounting systems developed by Pyatt and Roe (1977)[206].

2. The share of shipments transported on each mode can be determined endogenously by introducing mode K in equation 5.13 (assuming the cost function of mode k, c_a^k, is known) and by replacing X^{ijp} by X^{ijkp}

in corresponding equations. Work by Chon et al. (1983)[63] describes a model formulation and solution algorithm of this type in an urban context.

3. Use of commodity industry input-output coefficients as opposed to industry-industry coefficients would further enrich the results of the model, particularly in finding an efficient transportation system. Similar works of this nature can be found in Hewings (1982)[118].

4. Finally, the work of Bell, Hazell, and Slade (1982)[25] has provided a framework in which a single-region social accounting model (with an input-output component) could be linked with project appraisal analysis of the cost-benefit kind. An extension to a multiregional or interregional context would appear to offer many useful insights and draw together the framework provided in this paper with the concerns implicit in project selection and evaluation.

Chapter 6

Model Applications: Formulating Highway Investment Strategies

6.1 Introduction

Improvement of road conditions has become a major concern for many developing countries in the 1980s.[1] Roads in many developing countries have deteriorated in the past decade due to their bearing heavier traffic loads than was originally intended for them, and funds have been allocated for expanding rather than maintaining the network (World Bank 1983)[13]. Thus, formulating investment priorities for road improvement has become an important component in national planning strategy, because transportation is vital to nations' economies. Also, many countries face various problems from having allocated limited capital resources in the past to improving transportation facilities that have since deteriorated.

Based on models introduced in the previous chapter, this chapter introduces an extended model. The model is applied to Korea with the following purposes:

1. to shed light on issues of formulating investment priorities for road improvements in Korea using interzonal freight movement data with a transportation network of 174 nodes and 834 links;

[1] Portions of this chapter, in its first form, was written with Chang Ho Park and Jeong Hyun Rho, and was published in *Papers of Regional Science Association*, 1985, 57:91–105.

2. to compare changes in priorities for road improvements with alternative national growth policies.

For this study, Korea was divided into twenty-five zones (Figure 6.1) made up of Korea's sixteen official planning districts (*So-Kwon* in Korean) plus nine new districts, (that were made by dividing some districts into two or more, according to the level of economic activity).

6.2 The Model

Given input-output production relations (**A**) among $r = 1, \ldots, R$ sectors, and assuming that the total amount of each commodity to be exported from the nation is given (**E**), the total amount of each good to be produced (**X**) from the country as a whole is

$$\begin{aligned} \mathbf{X} &= \mathbf{AX} + \mathbf{E} \\ \mathbf{X} &= [\mathbf{I} - \mathbf{A}]^{-1}\mathbf{E} \end{aligned}$$

where **X** is an R by 1 vector, **A** is an R by R matrix and **E** is an r by 1 vector.

Further assume that the nation is subdivided into zones, $i = 1, \ldots, N$, and that export zones are designated, such as ports from which goods are exported at given cost (d_r^i). Once the transportation cost between zone i and zone j is given (c_r^{ij}), the cost minimization problem becomes one of finding the amounts of commodity r to be produced in each zone (X_r^i) and exported from each export zone (E_r^i) such that total transportation costs, export costs, and production costs are minimized subject to export constraints and the conservation of flow. The solution results in the identification of at most $2N - 1$ positive flows.

The cost-minimizing solution does not allow for shipments between two regions in both directions since, in principle, one of these shipments would be unnecessary or inefficient. A method is needed to modify the transportation problem formulation to provide a more realistic description of the commodity flow pattern. Erlander (1977)[85] proposed that the entropy function

$$-\sum_i \sum_j X_r^{ij} ln(X_r^{ij}) = S_r \tag{6.1}$$

be used to accomplish this objective. S_r can be interpreted as a scalar measure of the level of dispersion or crosshauling in an observed commodity

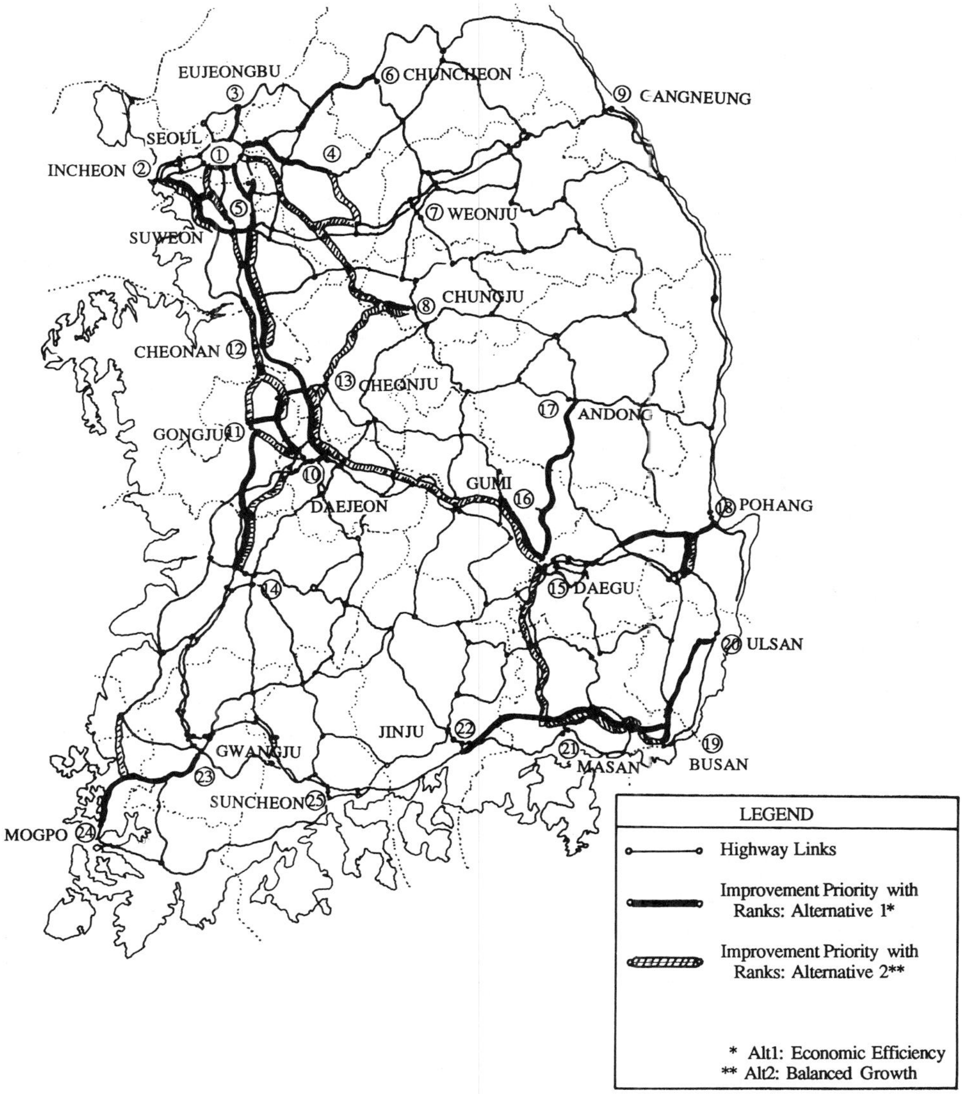

Figure 6.1: Priority Links for Improvements According to Alternative Growth Policies (Alt. 1: Economic Efficiency; Alt. 2: Balanced Growth)

flow matrix. This measure accounts for the observed pattern of crosshauling in both of the senses discussed above. By adding constraint (6.1) to the transportation problem, the solution is constrained to produce a flow matrix that has the same level of crosshauling as the observed matrix .

In the absence of origin-destination rate information for alternative routes as described earlier, further assume that each link of the network has a unit user cost, c_a, which is an increasing function of the annual flow, f_a :

$$c_a = c_a(f_a).$$

The behavior of the shipper in this situation may be generally depicted in terms of cost minimization. If link costs are fixed, the shipper will choose the path or sequence of links through the network that has the minimum total cost. If congested conditions exist in the network, then the transportation cost on each link is not fixed, but may be regarded as an increasing function of the flow on that link. Because any given link may serve many origin-destination pairs, the problem is highly interdependent. If shippers continue to seek their minimum cost paths, then a network equilibrium problem results.

The mathematical formulation of the network equilibrium problem can be formulated as follows:

$$P6.1)\ \min \quad \sum_a \int_0^{f_a} c_a(x)dx + \sum_i \sum_r p_r^i X_r^i + \sum_{i \in e} \sum_r d_r^i (E_r^i + I_r^i) \qquad (6.2)$$

$$\text{s.t.} \quad \sum_{i \in e} E_r^i \geq E_r, \sum_{i \in e} I_r^i \geq I_r \qquad \text{for all} \quad r \qquad (6.3)$$

$$I_r^i + \sum_j X_r^{ij} + X_r^i =$$

$$\sum_j X_r^{ij} + \sum_q a_{rq} X_q^i + E_r^i \quad \text{for all} \quad r \quad \text{and i} \qquad (6.4)$$

$$\sum_p X_r^{ijp} = X_r^{ij} \qquad \text{for all} \quad r, i, \text{and} \quad j \qquad (6.5)$$

$$f_a = \sum_r g_r \sum_i \sum_j \sum_p X_r^{ijp} \delta_a^{ijp} \qquad \text{for all} \quad a \qquad (6.6)$$

$$-\sum_i \sum_j (\sum_p X_r^{ijp}) \ln(\sum_p X_r^{ijp}) \geq S_r \qquad \text{for all} \quad r \quad (6.7)$$

$$X_r^{ijp} \geq \qquad \text{for all} \quad r, i, j, \quad \text{and} \quad p, \qquad (6.8)$$

where the endogenous variables are

E_r^i = amount of r exported from port i

I_r^i = amount of r imported through port i

X_r^i = amount of r produced at i

X_r^{ij} = amount of r shipped from i to j

X_r^{ijp} = amount of r shipped from i to j on path p

f_a = total flow on link a

and the exogenous variables are

E_r = total amount of r to be exported from the nation

I_r = total import quota for r

d_r^i = handling costs of r at port i

p_r^i = cost of producing a unit of r at i

g_r = road space occupancy required for shipping a unit of r

$$\delta_a^{ijp} = \begin{cases} 1, \text{if link } a \text{ is included in path } p \text{ from } i \text{ to } j \\ 0, \text{otherwise} \end{cases}$$

S_r = level of spatial interaction for the shipment of r

a_{rq} = amount of input q required per unit output of r.

The solution to this problem is often referred to as a "user-equilibrium" solution, as mentioned in Chapter 5. Unless link costs are constant, this solution does not minimize total transportation costs. The solution that does minimize total cost, referred to as the "system-optimal" solution, can be found by replacing the objective function with

$$\sum_a f_a c_a(f_a) = \sum_a \int_0^{f_a} m_a(x) dx \qquad (6.9)$$

where the marginal cost of link a is given by

$$m_a(x) = d[c_a(f_a) f_a]/df_a \qquad (6.10)$$

The minimization of equation (6.9) subject to equations (6.3) to (6.8) results in a nonlinear programming problem with a convex objective function. The problem becomes one of finding X_r^{ijp} given the cost function $c_a(x)$ and values of the exogenous variables.

Setting optimality conditions (see Appendix A) with respect to E_r^i, I_r^i, X_r^i, and X_r^{ijp} and solving for $\sum_p X_{r*}^{ijp}$, the following equation is obtained.

$$\sum_p X_{r*}^{ijp} = \exp \quad [\beta_r(\alpha_r^j - \sum_q \alpha_q^i(a_{rq}) - p_r^i - c_{r*}^{ij}]. \tag{6.11}$$

Volume for each link at optimum is then

$$f_a^* = \sum_r g^r \sum_i \sum_j \sum_p X_{r*}^{ijp} \alpha_a^{ijp}. \tag{6.12}$$

In equation (6.11), α_r and β_r are the Lagrange multipliers associated with equations (6.4) and (6.7), respectively.

Thus, α_r^j can be interpreted as the locational surplus arising from producing unit r at j and $\sum_q \alpha_q^i$ can be interpreted as the locational surplus arising from consuming r at i for the production of other goods. The equation (6.11) can be interpreted as follows: total commodity flow of r between i and j is positively related to α_r^j and inversely related to factors concerning consumption at $i(\sum_q \alpha_q^i)$, production cost at $i(p_r^i)$, and shipment costs (c_{r*}^{ij}) from i to j. Thus, equation (6.11) is the demand function for shipping r from i to j.

6.3 Solution Algorithm

The algorithm proposed by Evans (1973[88], 1976[87]) and later generalized by Chon et al. (1983)[63] was used to solve the model described above. The algorithm involves two solutions at each iteration: the main problem solution and the subproblem solution for determining the search direction for an improved main problem solution. The following are detailed descriptions of solutions: Let $f_a(n)$ be the flow of PCUs on link a in the main problem at iteration n; $k_a(n)$ be the flow in the subproblem at iteration n; $w^{ij}(n)$ be the flow volume between i and j in the main problem at iteration n; and $t^{ij}(n)$ be the flow volume in the subproblem at iteration n. Then the generalized Evans algorithm for solving the model presented in this paper consists of the following steps:

Step 1 Initialization: w^{ij} (1) and β are given from the observed data, set $f_a = 0$ and $n = 1$.

Step 2 Find minimum cost paths between i and j pairs by the following:

$$c^{ij}(n) = \sum_a c_a(f_a)\delta_a^{ijp}$$

where

$$c_a(f_a) = t_a[1 + 0.15(\frac{f_a}{s_a})^4], \tag{6.13}$$

$s_a =$ the capacity of link a, and

$t_a =$ unit shipment cost on link a when the link is not congested.

Step 3 Find the subproblem solution for $t^{ij}(n)$ by solving a doubly-constrained gravity model for each commodity r:

$$t_r^{ij}(n) = \gamma_r^i \cdot s_r^j \quad \exp \quad (-\beta_r c^{ij}) \tag{6.14}$$

where

$$\gamma_r^i = A_r^i O_r^i = A_r^i \sum_j t_r^{ij}(n) \tag{6.15}$$

$$s_r^j = B_r^j D_r^j = B_r^j \sum_i t_r^{ij}(n) \tag{6.16}$$

$$t_r^{ij} \geq 0. \tag{6.17}$$

Step 4 Assign $t_r^{ij}(n)$ to the minimum cost paths resulting in $k_a(n)$:

$$k_a(n) = \sum_r g_r \sum_i \sum_j \sum_p t_r^{ijp}\delta_a^{ijp}. \tag{6.18}$$

Step 5 Perform a line search in order to obtain a new main problem solution and to find the value $\lambda * \varepsilon[0,1]$ that minimizes:

$$\lambda * = \quad \min. \quad \{\sum_i \sum_j \sum_r \frac{1}{\beta} w_r^{ij}(\ln w_r^{ij} - 1) + \sum_a \int_0^{f_a} m_a(x)dx\} \tag{6.19}$$

$$0 \leq \lambda \leq 1$$

where

$$f_a(n+1) = (1 - \lambda*) f_a(n) + \lambda * k_a(n)$$
$$w_r^{ij}(n+1) = (1 - \lambda*) w_r^{ij}(n) + \lambda * t^{ij} - r(n)$$

Step 6 Perform convergence check:

$$\Delta \frac{\sum_i \sum_j w_r^{ij} c^{ij}}{\sum_i \sum_j w_r^{ij}} \quad \text{and/or}$$
$$\Delta \quad \text{(objective function value) and/or}$$
$$\Delta \sum_i \sum_j c^{ij} \sum_r (w_r^{ij} - t_r^{ij}).$$

If these are satisfied, stop. Otherwise, go to step 2 and $n = n + 1$.

Step 7 Formulate a priority index (I_a) for road improvements for all links:

$$I_a = f_a \cdot c_a(f_a)/l_a = \int_0^{f_a} m_a(x)dx/l_a \tag{6.20}$$

where l_a = distance of link a.

6.4 Setting Investment Priorities

Gartner (1980)[106] presented economic rationales of the solutions associated with system optimization once the demand functions $X^{ij} = f^{ij}(c^{ij})$ and travel performance functions $c_a(f_a)$) are known. If there is only one link that connects i and j, the area OMEF in Figure 6.2 is equal to the optimum objective function value associated with the link. The same results can be obtained by maximizing the area FER, which represents "system surplus." Thus, the area OMEF can be interpreted as travel disutility; the objective of system optimization is, then, the reduction of aggregate travel disutility represented by the area OMEF.

The economic rationales stated above hold for situations where two or more links connect O/D pairs (Figure 6.2). The areas OMEF and ONQC

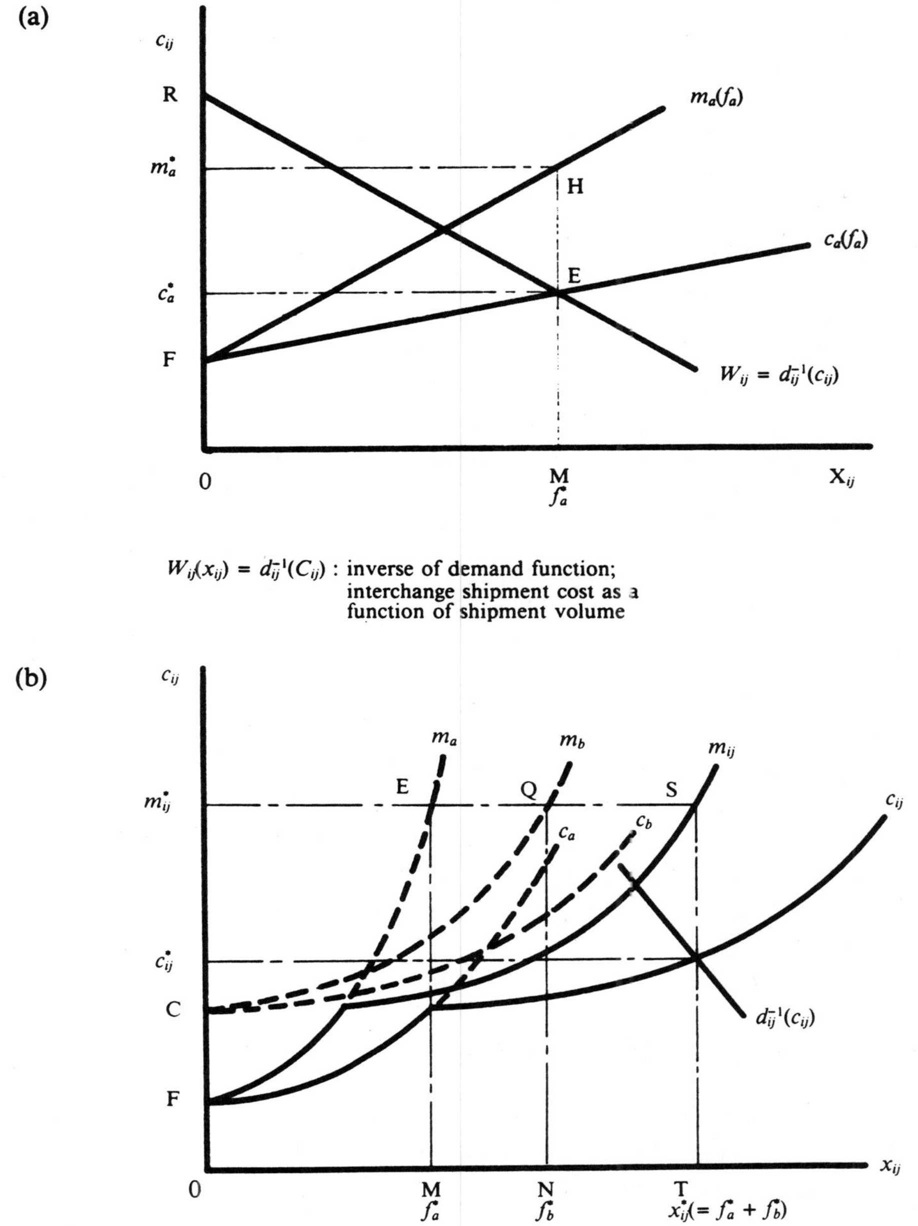

Figure 6.2: Relationship between Travel Cost and Travel Volume: Case (a) in which one link exists between i and j; case (b) in which two or more links exist between i and j

represent travel disutility for taking link a and b, respectively, at equilibrium. The total travel disutility for making a trip from i to j at equilibrium is represented by the area OTSF, which equals OMEF and ONQC.

The travel disutility thus identified can shed light on the issue of setting priorities for road improvements. The link that shows the highest travel disutility should receive the highest priority for the improvements, since improvements will reduce total system costs most and will increase system surplus most at equilibrium.

The above analysis concerns a static situation only. If an investment is performed on one road, it will then cause traffic to decrease or increase on other roads and, therefore, have an influence on the investment decisions to be made for those roads. The application of the proposed model will still be useful for the analysis of dynamic road investment planning because it provides valuable information for generating alternative networks by identifying locations of "bottlenecks." Dynamic road investment analyses typically assume that alternative networks are known a priori (Bergendahl 1969)[30].

Despite its formulation as a static model, travel disutility thus identified is the result of interaction between (1) public sectors represented by the agent who sets export-import targets and provides efficient transportation systems and (2) private sectors represented by the agent who chooses minimum costs of production ($p_r^i X_r^i$) and shipment (X_r^{ijp}).

The model assumes the export-import targets (E_r, I_r) as given, but zonal production (X_r^i) and consumption ($\sum_q a_{rq} X_q^i$) can either be given or determined by the model. Thus, any known changes in export-import targets or zonal production/consumption can be introduced in order to find the magnitude of travel disutility associated with each link of the road network, which in turn provides valuable information for formulating priorities for road improvements.

6.5 Application to Korean Networks I

To test the empirical performance of the model, a simplified data set from Korea is used. The original interregional shipment data collected in 1978 for 100 origins and 100 destinations for eighteen commodities were grouped into seven regions and one combined commodity. Transportation nodes that connect seven regions were also simplified for this analysis (Figure 6.3).

It is important to note here that the model is intended to provide investment priorities by identifying the magnitude of "bottlenecks" or "disutility,"

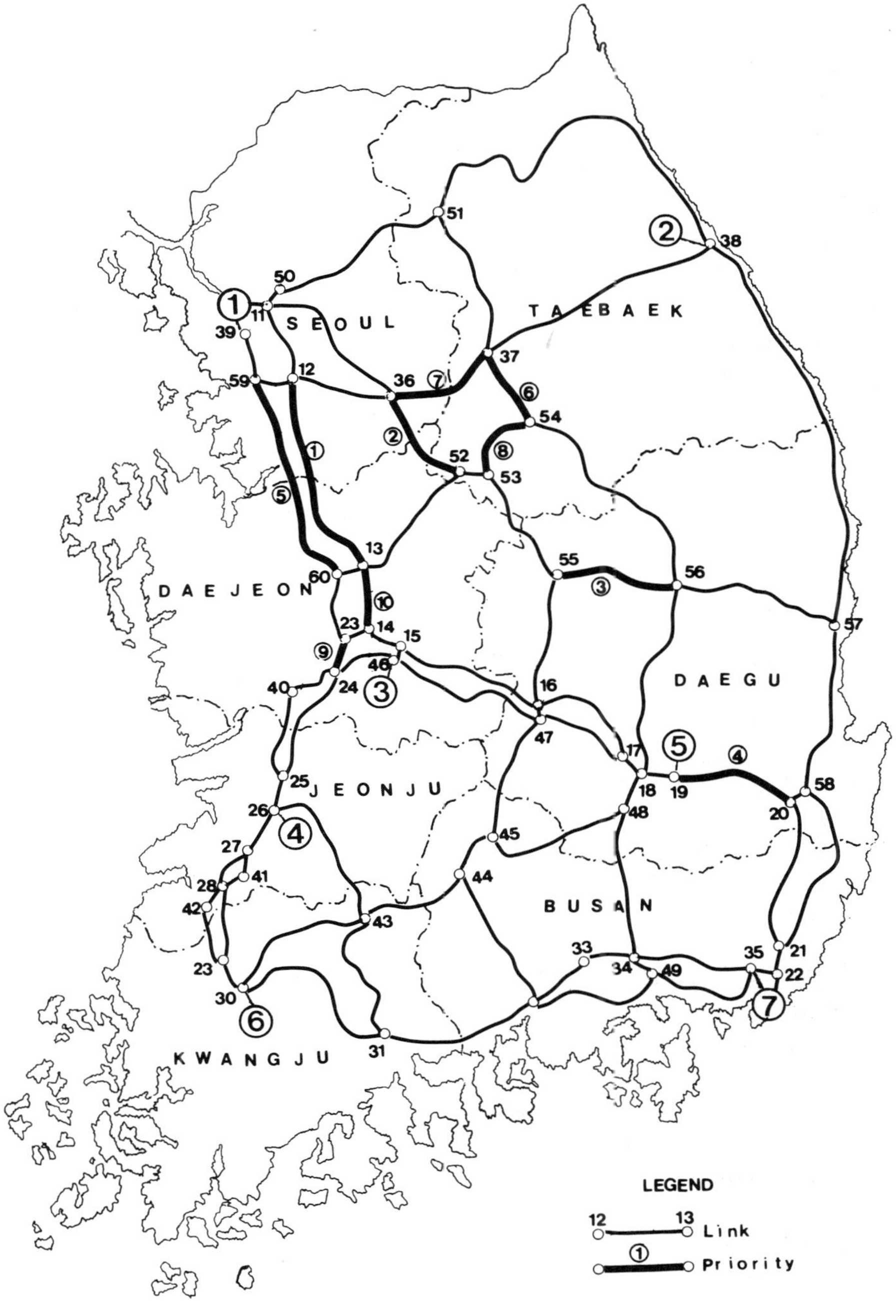

Figure 6.3: Highway map for 7 Zones in Korea

both of which are based on aggregate estimates of interregional shipment patterns. Although based in part on behavioral postulates, the model is clearly not intended to present the individual behavior of shippers or carriers. Thus, the testing of the model in this section will only be concerned with comparison between observed shipment patterns and aggregated model estimates upon which investment priorities will be identified.

Given the 1978 interregional shipment data, the export-import volume at each port (E_r^i, I_r^i) and the production (X_r^i) and consumption in each region ($\sum_q a_{rq} X_q^i$) are known and fixed. Thus, the model presented above can further be simplified by minimizing the objective function (6.9), subject to constraints (6.5) through (6.8) and

$$\sum_j X_r^{ij} = O_r^i \quad \text{for all} \quad i \quad \text{and} \quad r \tag{6.21}$$

$$\sum_i X_r^{ij} = D_r^j \quad \text{for all} \quad j \quad \text{and} \quad r, \tag{6.22}$$

where

O_r^i = total amount of r originating from i (observed) and

D_r^j = total amount of r arriving at j (observed).

The following assumptions are made for the numerical analysis:

1. All *intrazonal* shipment is ignored for the sake of simplification. Losses of information may be minimized if one divides the area into smaller zones.

2. Units in tons were converted to the number of trucks required for shipment, which in turn were converted to passenger car equivalent units (PCU).

3. The capacity of each link was based on a function of physical conditions of the road segments, intrazonal shipment, passenger car trips, area type, facility type, and number of lanes.

4. The congestion function used here was the same as the function of the Bureau of Public Roads. This function may have to be estimated separately for each country for better approximation.

   It is given by:

$$c_a(f_a) = t_a[1 + 0.15(\frac{f_a}{s_a})^4] \tag{6.23}$$

where

$c_a(f_a) =$ the generalized cost of shipment on link a with flow volume f_a

$f_a =$ volume on link a

$s_a =$ capacity of link a, and

$t_a =$ unit shipment cost on link a when the link is not congested.

The solution algorithm for minimizing equation (6.9) subject to equations (6.5) through (6.8), (6.21), and (6.22) is described well by Evans (1976)[87] and Chon et al. (1983)[63]. Using the solution algorithm as modified by Chon et al. (1983)[63], estimates of the interregional shipment patterns are compared with the observed shipments in Table 6.1. The PCUs have been converted back to tonnage to facilitate the comparison of observed shipment patterns and model estimates. This comparison is highlighted in Figure 6.4.

The location of the ten links with highest travel disutility are shown in Figure 6.3. Disutility values range from 256.2 for the link with highest priority to 10.8 for the link with the tenth highest priority. The results show that the highest investment priority should be given to improving roads that connect Seoul and Daejeon.

The findings of the analysis should not be interpreted as the final recommendation for investment priorities in Korea. Representations of regions, nodes, and the network are too schematic to draw any reasonable policy recommendations. Nevertheless, it is demonstrated that the model can shed light on the issue of identifying investment priorities for road improvement, given origin and destination amounts of shipments.

6.6 Application to Korean Networks II

Table 6.2 presents Korea's shipment pattern for twenty-five origins and destinations for freight movements. Table 6.3 shows origin-destination trips for passengers. Both tables were constructed based on a 1978 survey that covered eighteen commodities plus passengers for 100 origins and 100 destinations. Units in these tables are tons/day for freight and PCUs/day for passengers, as described in detail in Appendix B.

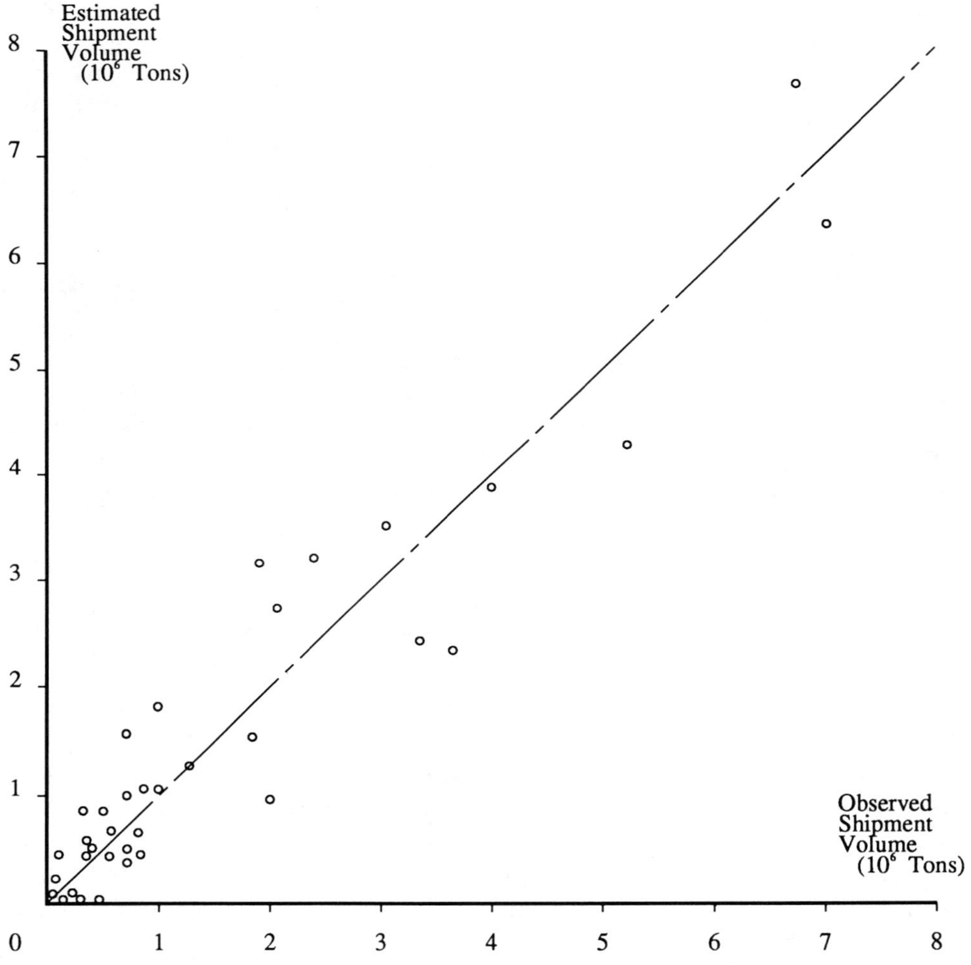

Figure 6.4: Comparison between Observed and Estimated Shipment Patterns: Total

O/D	1	2	3	4	5	6	7	TOTAL
1	100,879	2,216	4,139	1,365	2,102	6,997	1,840	119,538
	(100,879)	(2,832)	3,935)	1,303)	(3,252)	(5,847)	(1,493)	(119,541)
2	2,497	6,680	365	15	270	139	15	9,981
	(3,219)	(6,680)	(11)	(4)	(11)	(44)	(4)	(9,973)
3	3,570	245	13,155	719	442	650	343	19,125
	(2,413)	(7)	(13,155)	(507)	(876)	(1,580)	(584)	(19,122)
4	847	18	840	3,165	135	237	745	5,987
	(708)	(4)	(540)	(3,165)	(256)	(883)	(431)	(5,987)
5	3,128	252	799	164	22,488	5,311	310	32,452
	(3,471)	(26)	(1,099)	(416)	(22,488)	(4,318)	(635)	(32,453)
6	6,840	117	661	296	3,803	47,614	986	60,317
	(7,723)	(58)	(993)	(551)	(2,278)	(47,614)	(1,099)	(60,316)
7	2,994	29	350	668	336	1,066	10,063	14,516
	(1,062)	(7)	(650)	(453)	(438)	(1,843)	(10,063)	(14516)
TOTAL	119,765	9,557	20,309	6,392	29,576	62,014	14,302	261,915
	(119,275)	(9,614)	(20,383)	(6,399)	(29,599)	(62,129)	(14,309)	(261,980)

Note : Orginal data were collectd for one hundred origins and one hundred destinations, and for eighteen commodities by a research team in Regional Development Research Institute (RDRI/KIST) in 1978. Estimated shipments are given in parentheses. Units are in thousands of tons per year.

Table 6.1: Estimated and Observed Interregional Shipment Patterns for 7 O/D Zones:1978

The dispersion measures S_r were based on Tables 6.2 and 6.3 such that

$$\begin{aligned} S_{\text{freight}} &= -0.206E10^3 \\ S_{\text{passenger}} &= -0.847E10^3 \end{aligned}$$

For the numerical analysis, these values of S_r are used to ensure that the level of dispersion in 1978 should at least be maintained. The network, which consists of 174 nodes and 834 links, is shown in Figure 6.1.

The model represented by minimizing equation (6.9), subject to equations (6.5) through (6.8), and equation (6.21) and (6.22) is applied to Korea using data from Tables 6.2 and 6.3, with the transportation network consisting of 174 nodes and 834 links. The distribution results are shown in Table 6.4 for freight and in Table 6.5 for passenger flow.

Figure 6.5 shows a comparison between observed shipment patterns and model results for freight movements. The correlation coefficient between the two is 0.910. Figure 6.6 shows the same comparison for passengers, with a correlation coefficient of 0.931.

Equilibrium assignment procedures as outlined in the previous section resulted in a priority index for road improvements in Korea (Table 6.6). As

O/D	1	2	3	4	5	6	7	8	9	10	11	12	13
1	0	21600	4040	2600	11600	1870	988	534	806	2580	696	467	2170
2	31000	0	390	758	7980	155	252	341	116	1930	95	1	1020
3	4390	270	0	88	164	15	3	5	1	24	0	9	22
4	4110	1610	139	0	941	57	153	44	9	0	0	30	5
5	8940	3760	874	749	0	74	335	67	251	521	53	275	451
6	1800	26	44	47	24	0	246	11	209	37	0	4	0
7	1750	108	4	260	245	478	0	397	510	72	24	0	228
8	1050	0	1	28	54	21	228	0	23	474	1	100	27
9	1330	58	0	5	9	413	427	39	0	9	0	14	14
10	2570	881	34	16	502	32	101	226	0	0	1340	925	1190
11	1020	30	0	0	39	0	3	2	0	1680	0	347	173
12	556	219	0	61	77	11	4	245	0	860	458	0	57
13	2730	397	0	3	649	11	1	14	17	999	665	135	0
14	1990	139	12	29	158	0	11	34	8	1040	816	180	261
15	2560	550	132	3	277	11	19	31	94	1120	30	38	345
16	371	37	2	4	25	3	0	7	3	244	0	32	44
17	368	2	0	9	2	5	19	216	136	33	5	90	0
18	3630	360	13	0	237	17	2	5	123	160	18	6	31
19	12300	1060	152	17	469	27	9	14	39	519	26	230	566
20	1800	241	2	0	172	0	0	0	10	263	0	16	40
21	1810	36	0	0	105	1	5	6	163	59	0	2	26
22	587	4	3	0	10	7	0	6	30	30	15	3	18
23	2390	276	4	53	204	0	16	35	10	372	27	41	41
24	1130	43	1	0	13	0	0	0	0	275	0	4	0
25	1350	26	3	0	4	0	1	0	16	100	103	1	0
TTL	91532	31732	5848	4729	23960	3207	2822	2279	2575	13402	4372	3941	6730

O/D	14	15	16	17	18	19	20	21	22	23	24	25	TOTAL
1	2990	2610	460	208	1020	12400	1790	1440	238	2290	508	765	76670
2	511	520	27	22	206	1450	50	148	127	347	67	530	49032
3	37	13	0	0	2	115	0	0	1	1	8	0	5167
4	0	0	7	27	0	0	0	0	2	49	32	28	7242
5	195	305	129	175	36	1110	84	254	6	328	29	60	19062
6	0	41	5	6	0	19	0	8	10	1	0	0	2538
7	5	8	0	29	0	57	0	12	0	0	0	0	4186
8	29	25	2	121	3	124	4	2	3	1	0	1	2322
9	1	203	0	137	162	34	3	81	21	33	0	0	2993
10	1250	388	473	25	90	573	239	75	232	410	112	314	11998
11	381	19	5	0	0	38	0	0	2	45	0	1	3785
12	18	37	32	10	17	150	0	10	4	2	11	0	2837
13	321	85	27	6	0	430	0	12	10	34	12	1	6558
14	0	277	21	49	21	565	6	14	63	1270	274	501	7738
15	342	0	4270	1550	2780	2600	485	1330	488	292	63	110	19519
16	7	5580	0	291	67	1210	17	10	51	7	4	7	8024
17	22	1650	526	0	313	326	6	9	50	157	6	8	3957
18	83	2330	467	294	0	4410	2490	1040	31	43	26	124	15938
19	614	3660	616	176	2600	0	5890	11400	1510	993	85	575	43547
20	74	670	68	12	704	8490	0	508	32	91	0	78	13270
21	38	993	31	5	75	13400	504	0	1190	214	88	138	18888
22	80	607	134	27	40	2350	39	2230	0	59	60	319	6659
23	994	462	18	106	65	1050	12	154	157	0	3020	1800	11307
24	317	115	15	0	8	303	0	48	23	2480	0	299	5073
25	519	109	2	7	12	784	76	149	169	1370	422	0	5224
TTL	8829	20707	7334	3283	8219	51988	11694	18933	4419	10517	4826	5658	353535

Table 6.2: Observed Interzonal Freight Movement in 1978 (PCU/Day)

O/D	1	2	3	4	5	6	7	8	9	10	11	12	13
1	0	31533	18400	23267	33200	1553	1493	1100	2220	2373	2200	2967	6800
2	32467	0	48	720	9667	767	4	74	30	404	33	329	306
3	19667	0	0	7467	20	442	26	27	26	233	0	92	0
4	23400	1480	6800	0	5420	1933	940	1900	227	47	0	1567	11
5	36067	5180	129	5287	0	152	147	159	4	401	987	3300	5887
6	1593	507	303	1820	67	0	1787	41	1507	30	0	0	0
7	1527	4	4	767	126	1833	0	3220	2227	0	0	0	4
8	1040	181	14	1593	99	24	3493	0	62	443	35	3000	0
9	1913	0	0	241	56	1260	2267	63	0	7	0	0	67
10	2373	234	0	15	394	0	0	181	11	0	11467	2767	3747
11	2333	0	0	0	707	0	0	126	0	10867	0	1827	8400
12	3647	30	11	1247	2233	0	44	3980	0	4347	867	0	1593
13	7333	243	666	4	5587	0	0	7	26	2967	7733	1740	0
14	3233	85	0	21	489	11	0	2	26	6247	5927	76	1333
15	1753	81	19	0	48	4	30	12	67	654	0	59	48
16	614	0	0	12	30	0	4	31	0	1793	0	1167	19
17	214	0	185	211	0	0	133	907	33	30	0	286	11
18	633	74	0	7	26	0	0	0	1620	102	4	0	0
19	1167	30	167	0	0	0	0	4	30	190	0	4	11
20	707	315	0	0	85	0	0	0	0	394	4	11	48
21	656	222	4	4	59	0	0	4	11	96	19	70	0
22	133	0	139	26	193	0	11	0	0	26	0	22	0
23	2113	59	26	0	59	0	0	19	4	148	4	0	26
24	168	0	0	0	26	0	0	0	4	43	0	4	133
25	353	0	0	0	7	0	0	0	0	67	59	11	4
TTL	145104	40258	26913	42706	58597	7979	10379	11856	8132	31910	29337	19298	28447

O/D	14	15	16	17	18	19	20	21	22	23	24	25	TOTAL
1	2787	1527	599	223	537	1013	498	604	129	2020	157	338	137538
2	37	48	15	0	26	15	0	0	30	7	0	0	45026
3	0	48	4	115	0	0	0	15	260	0	0	0	28441
4	2	44	42	260	0	0	0	19	4	0	0	0	44096
5	283	52	15	4	11	7	0	30	525	0	19	7	58652
6	7	4	0	0	26	0	0	7	0	0	0	0	7697
7	0	26	0	107	11	0	0	11	0	0	0	0	9866
8	67	7	66	1007	0	19	30	0	4	0	4	15	11202
9	11	52	0	26	1413	4	30	41	0	0	0	4	7453
10	6600	807	1853	68	91	161	324	107	26	166	47	63	31501
11	6413	4	0	0	0	4	7	4	0	15	0	0	30706
12	105	26	1347	383	0	7	0	59	7	0	0	11	19944
13	1213	4	11	15	0	41	0	0	7	33	126	0	27757
14	0	294	321	5	4	151	4	20	1580	7133	260	3860	31081
15	259	0	16800	8200	11800	1740	648	4080	4920	116	152	133	51623
16	319	14667	0	5860	19	115	19	73	2313	10	0	79	27142
17	1	6560	5893	0	1973	41	0	4	22	33	0	0	16538
8	11	8133	41	2333	0	2853	1993	44	19	17	0	4	17915
19	198	1753	70	67	3200	0	12667	14600	4647	393	4	563	39761
0	89	747	7	4	2047	12400	0	507	127	137	0	19	17647
21	134	3767	125	0	37	13000	588	0	11267	232	26	246	30565
2	1833	4187	2313	56	4	5013	99	9867	0	297	53	3020	27291
23	7533	137	23	0	17	415	19	181	329	0	19600	9200	39912
64	235	93	4	0	0	7	0	41	57	16600	0	2387	19801
25	3960	169	89	7	0	628	11	309	3380	6933	3087	0	19075
TTL	32099	43153	29639	18738	21216	37634	16936	30622	29653	34143	23534	19947	798229

Table 6.3: Observed Interzonal Passenger movement in 1978 (PCU/Day)

O/D	1	2	3	4	5	6	7	8	9	10	11	12	13
1	0	28500	4590	3010	19000	2140	1600	1170	1210	3100	1070	1710	2100
2	42100	0	776	339	2830	251	180	132	136	461	159	255	312
3	4130	478	0	49	188	40	26	19	19	44	15	24	33
4	4640	358	83	0	291	206	230	116	211	212	66	106	204
5	15300	1540	168	159	0	79	86	63	64	328	113	181	223
6	1770	137	37	111	81	0	80	29	73	39	13	22	36
7	2430	187	43	226	154	146	0	101	188	128	35	57	123
8	1210	93	22	78	77	37	69	0	47	132	36	34	127
9	1500	116	27	170	95	110	154	57	0	72	22	35	70
10	2770	279	45	118	365	44	72	109	50	0	640	329	862
11	881	89	14	37	116	14	20	29	15	615	0	121	230
12	1230	124	20	52	162	20	28	23	21	264	105	0	179
13	2040	205	34	123	269	40	75	114	52	935	254	242	0
14	1120	113	18	48	148	18	29	44	20	1140	410	134	350
15	1380	140	23	59	183	22	41	60	49	1340	284	165	431
16	749	76	12	32	99	13	24	35	16	725	154	89	233
17	529	41	9	34	34	19	37	46	40	241	51	30	85
18	504	51	8	21	66	10	18	22	160	487	103	60	157
19	1360	137	22	58	179	22	40	59	156	1310	279	162	423
20	240	24	4	10	32	4	7	10	35	232	49	29	75
21	372	38	6	16	49	6	11	16	17	360	76	44	116
22	117	12	2	5	15	2	3	5	5	113	24	14	36
23	806	81	13	34	106	13	21	32	14	814	293	96	251
24	297	30	5	13	39	5	8	12	5	300	108	35	92
25	129	13	2	5	17	2	3	5	2	131	47	15	40
TTL	87604	32861	5985	4806	24595	3260	2860	2307	2607	13523	4407	3988	6788

O/D	14	15	16	17	18	19	20	21	22	23	24	25	TOTAL
1	1290	1410	681	436	260	1500	215	388	83	770	258	143	76634
2	191	210	101	49	39	223	32	58	12	115	38	21	49021
3	18	20	10	7	4	21	3	5	1	11	4	2	5170
4	88	97	47	43	18	102	15	27	6	53	18	10	7245
5	136	150	72	23	28	159	23	41	9	82	27	15	19068
6	16	19	10	13	5	20	3	5	1	10	3	2	2535
7	53	66	35	46	15	70	10	18	4	32	11	6	4184
8	55	66	35	39	12	70	10	18	4	33	11	6	2320
9	30	66	20	41	108	211	41	23	4	18	6	3	2998
10	1160	1220	589	168	225	1290	186	336	72	694	232	129	11984
11	401	248	120	34	46	263	38	68	15	240	80	45	3780
12	110	120	58	17	22	128	18	33	7	66	22	12	2840
13	388	426	205	65	79	452	65	117	25	233	78	43	6558
14	0	439	212	61	81	623	67	162	94	1530	514	354	7729
15	491	0	1650	591	1230	7290	1020	1890	372	356	138	303	19507
16	266	1850	0	180	341	1960	282	508	100	159	53	81	8037
17	88	796	210	0	349	832	131	216	43	53	18	35	3965
18	178	2430	599	527	0	7740	1490	840	149	143	56	122	15941
19	650	6990	1620	572	3880	0	7290	11700	2090	2000	774	1700	43472
20	85	1160	286	115	866	9020	0	640	114	109	42	93	13281
21	178	1850	443	157	433	12300	571	0	572	547	212	465	18855
22	154	540	129	46	115	3270	151	849	0	472	183	401	6663
23	1840	377	152	43	80	2280	106	592	344	0	1620	1300	11309
24	676	116	56	16	25	702	33	182	106	1780	0	435	5074
25	364	276	66	23	59	1670	77	434	252	1120	473	0	5228
TTL	8907	20943	7404	3312	8317	52197	11876	19151	4479	10624	4871	5726	353397

Table 6.4: Estimated Interzonal Freight Movement (PCU/Day)

O/D	1	2	3	4	5	6	7	8	9	10	11	12	13
1	0	34400	22800	17267	51133	2187	1233	2140	659	579	687	2587	1280
2	37867	0	2200	650	3667	91	47	81	25	42	49	185	91
3	23733	2133	0	973	940	177	69	121	37	24	28	106	69
4	16600	583	900	0	1367	4080	5780	4260	4853	491	467	1760	2427
5	50000	3260	880	1533	0	85	114	197	61	259	307	1160	573
6	2047	72	155	3940	81	0	565	203	475	11	13	50	50
7	1113	39	60	5413	95	549	0	927	1113	46	32	122	228
8	1827	64	99	3780	155	185	867	0	371	403	275	313	1987
9	571	20	31	4373	49	443	1073	384	0	19	17	62	94
10	501	33	22	396	229	11	39	367	17	0	8600	2353	6667
11	572	37	25	447	262	13	33	278	13	9133	0	3727	5067
12	2367	154	103	1847	1087	54	137	302	73	2493	4093	0	5507
13	1140	74	53	2020	521	43	201	1867	86	6867	4713	5360	0
14	137	9	6	109	63	3	11	101	5	5133	6733	647	1833
15	67	4	3	53	31	2	7	61	11	2273	880	317	900
16	69	4	3	55	32	2	9	74	4	2327	900	325	920
17	95	3	5	196	8	14	72	420	91	561	217	78	273
18	4	0	0	4	2	0	1	4	115	150	58	21	59
19	3	0	0	2	1	0	0	3	7	95	37	13	37
20	2	0	0	1	1	0	0	2	8	63	24	9	25
21	4	0	0	3	2	0	0	3	1	119	46	17	47
22	2	0	0	2	1	0	0	2	1	81	31	11	32
23	20	1	1	16	9	0	2	15	1	753	993	95	270
24	6	0	0	5	3	0	0	4	0	223	294	28	80
25	1	0	0	1	1	0	0	1	0	49	64	6	17
TTL	138747	40893	27347	43086	59738	7939	10261	11814	8035	32195	29561	19352	28533

O/D	14	15	16	17	18	19	20	21	22	23	24	25	TOTAL
1	166	72	86	149	6	4	2	5	4	23	8	2	137478
2	12	5	6	6	0	0	0	0	0	2	1	0	45026
3	7	3	4	8	0	0	0	0	0	1	0	0	28435
4	141	61	73	297	5	3	2	4	3	19	6	2	44182
5	74	32	39	14	3	2	1	2	2	10	3	1	58611
6	3	2	2	21	0	0	0	0	0	0	0	0	7691
7	13	8	11	103	1	0	0	0	0	2	1	0	9878
8	115	62	89	564	5	3	2	4	3	16	5	1	11199
9	5	12	5	125	159	7	9	1	1	1	0	0	7462
10	5313	2093	2513	680	183	103	67	134	104	727	242	61	31456
11	7333	853	1020	275	74	42	27	54	42	1007	335	85	30761
12	713	309	371	100	27	15	10	20	15	98	33	8	19937
13	1973	853	1027	347	74	42	27	55	42	269	90	23	27766
14	0	435	521	141	38	42	14	54	419	9933	3307	1380	31074
15	489	0	17933	8200	6120	3673	2260	4813	3047	103	48	292	51588
16	500	15333	0	2227	1333	747	493	980	621	69	23	60	27107
17	121	6733	2000	0	4327	319	257	418	265	16	5	25	16521
18	32	4647	1187	4287	0	2873	3693	504	253	9	4	24	17932
19	41	3387	747	329	3747	0	9200	14067	7067	240	112	673	39807
20	14	1953	499	295	4287	9467	0	624	313	11	5	30	17633
21	51	3953	940	416	597	13467	649	0	8933	303	142	853	30548
22	355	2253	538	237	272	6140	295	8067	0	2100	987	5920	27328
23	10733	97	77	21	12	264	13	346	2660	0	14800	8733	39933
24	3180	19	23	6	2	52	2	68	521	13200	0	2100	19818
25	1120	207	50	22	25	565	27	740	5693	6647	3813	0	19050
TTL	32506	43384	29760	18870	21298	37828	17053	30961	30009	34805	23971	20275	798220

Table 6.5: Estimated Interzonal Passenger Movement (PCU/Day)

```
SCATTERGRAM OF    (DOWN) PRDF     PREDICTED FREIGHT INTERZONAL TRIP
                (ACROSS) OBSF     OBSERVED FREIGHT INTERZONAL TRIP

              3445.00  10335.00  17225.00  24115.00  31005.00
         .+----+----+----+----+----+----+----+----+----+.
42102.00 +               I              I              *+      42102.00
         I               I              I               I
         I               I              I               I
         I               I              I               I
         I               I              I               I
37424.00 +               I              I               +      37424.00
         I               I              I               I
         I               I              I               I
         I               I              I               I
         I               I              I               I
32746.00 +               I              I               +      32746.00
         I               I              I               I
         I               I              I               I
         I               I              I               I
         I               I              I               I
28068.00 +-------------------------------*--------------+      28068.00
         I               I              I               I
         I               I              I               I
         I               I              I               I
         I               I              I               I
23390.00 +               I              I               +      23390.00
         I               I              I               I
         I               I              I               I
         I               I              I               I
         I               I              I               I
18712.00 +               I *            I               +      18712.00
         I               I              I               I
         I               I              I               I
         I               I              I               I
         I             * I              I               I
14034.00 +----------------------------------------------+      14034.00
         I               I              I               I
         I               I * *          I               I
         I               I              I               I
         I               I              I               I
 9356.00 +            *  I              I               +       9356.00
         I               I              I               I
         I    * *  *     I              I               I
         I     *         I              I               I
         I               I              I               I
 4678.00 +      2        I              I               +       4678.00
         I    * *        I              I               I
         I  33        *  I              I               I
         I 8635** *      I  *           I               I
         I99953*         I  *           I               I
       0 +9974           I              I               +             0
         .+----+----+----+----+----+----+----+----+----+.
           0    6890.00  13780.00  20670.00  27560.00

STATISTICS..

  CORRELATION (R)-           .90988     R SQUARED        -        .82788
  SIGNIFICANCE R -           .00001     STD ERR OF EST   -    1029.92505
  INTERCEPT (A)  -        -57.33002     STD ERROR OF A   -      42.73914
  SIGNIFICANCE A -           .09014     SLOPE (B)        -       1.10092
  STD ERROR OF B -           .02011     SIGNIFICANCE B   -        .00001
```

Figure 6.5: Comparison between Observed and Estimated Shipment Patterns: Freight

```
SCATTERGRAM OF    (DOWN)  PRDP     PREDICTED PASSENGER INTERZONAL TRIP
                  (ACROSS) OBSP    OBSERVED PASSENGER INTERZONAL TRIP

                    4008.00  12024.00  20040.00  28056.00  36072.00
                .+----+----+----+----+----+----+----+----+----+.
   51138.00     +               I               I          *    +   51138.00
                I               I               I              *I
                I               I               I               I
                I               I               I               I
                I               I               I               I
   45456.00     +               I               I               +   45456.00
                I               I               I               I
                I               I               I               I
                I               I               I               I
                I               I               I               I
   39774.00     +               I               I               +   39774.00
                I               I               I               I
                I               I               I          *    I
                I               I               I               I
                I               I               I               I
   34092.00     +---------------------------------------*-------+   34092.00
                I               I               I               I
                I               I               I               I
                I               I               I               I
                I               I               I               I
   28410.00     +               I               I               +   28410.00
                I               I               I               I
                I               I               I               I
                I               I               I               I
                I               I          *    I               I
   22728.00     +               I        *      I               +   22728.00
                I               I               I               I
                I               I               I               I
                I               I               I               I
                I               I      *        I               I
   17046.00     +------------------------------2----------------+   17046.00
                I               I               I               I
                I               I  *     *      I               I
                I               I* *   *        I               I
                I               I               I               I
   11364.00     +               I               I               +   11364.00
                I         2     I               I               I
                I           *  3**              I               I
                I          * *  I               I               I
                I    ****2*     I               I               I
    5682.00     + 22 2 * 2      *               I               +    5682.00
                I2*32 *    3    I               I               I
                I2 6 3*2     *  I               I               I
                I8682*3 2       I               I               I
                I944832 322     I               I               I
          0     +99952* *       I               I               +          0
                .+----+----+----+----+----+----+----+----+----+.
                  0     8016.00  16032.00  24048.00  32064.00

STATISTICS..

  CORRELATION (R)-          .93125      R SQUARED        -          .86723
  SIGNIFICANCE R -          .00001      STD ERR OF EST   -      1586.75843
  INTERCEPT (A)  -       -51.28992      STD ERROR OF A   -        66.79904
  SIGNIFICANCE A -          .22144      SLOPE (B)        -         1.03990
  STD ERROR OF B -          .01630      SIGNIFICANCE B   -          .00001
```

Figure 6.6: Comparison between Observed and Estimated Shipment Patterns: Passenger

described previously, the levels of disutility shown in the fourth column in Table 6.6 were estimated based on equation (6.20). The disutility level is the sum of the area under the marginal cost curve of link a divided by the distance of the link, representing a unit level of disutility, or "bottlenecks" in the system (Kim, Park, and Rho 1985)[141].

Results shown in Table 6.6 should be interpreted with caution for several reasons. First, the results reflect the origin-destination survey of 1978, which undoubtedly revealed shipment patterns quite different from those of today. Second, the results are very sensitive to the capacities of each link. It is extremely difficult to represent accurate capacities between nodes, because not all existing links can be represented. Moreover, unlike urban networks, where well-defined peak hours exist, no common delivery time can be found in interregional freight shipment patterns. This leads to additional difficulties in defining the capacity of a link. The capacity, usually defined as volume/day, would vary widely depending on the hour of day used in determining road capacity. Consequently, the average link cost and delay functions, such as the one shown in equation (6.13), also differ significantly depending on the definition of capacity of a link. For example, a link may be interpreted to be congested if shipment is allowed for only eight hours per day but may be uncongested if shipment is allowed all day. The alternative definition of capacity affects estimated shipment costs, which in turn affect the results of mode choice, route choice, and, ultimately, the priorities for national road improvements.

Nevertheless, the priority index (Table 6.6) seems to reflect bottlenecks in Korea's road network in the late 1970s. Furthermore, it is demonstrated that, given origin and destinational amounts of shipments and dispersion factor (S_r) that reflect national development policies, the model can shed light on the issue of identifying investment priorities for road improvement. The results shown above reflect the level of interzonal economic interactions in 1978, which in turn reflect the growth policy of the pre-1978 period. It is often quoted that the growth policy of this period was labeled "Export First" or "Efficiency Priority" (Choe and Kim 1985)[61].

The next section analyzes the changes in priorities for road improvement with alternative national growth policy.

Priority	Highway No.*	Link			Level of Disutility
1	N 42	SUWON	-	BANWOL	.112E+03
2	N 42	INCHEON	-	SIHEUNG	.994E+02
3	N 42	SIHEUNG	-	BANWOL	.992E+02
4	N 42	SUWON	-	SINGAL	.605E+02
5	N 6	DONONG	-	DEOGSO	.400E+02
6	N 6	DEOGSO	-	YANGSU	.230E+02
7	N 6	YANGSU	-	YANGPYEONG	.225E+02
8	N 1	SUWON			.127E+02
9	E 1	SEONGNAM	-	SINGAL	.123E+02
10	E 1	SEOUL	-	SEONGNAM	.110E+02
11	E 1	SINGAL	-	SUWON	.101E+02
12	N 7	BUSAN	-	DONGRAE	.996E+01
13	N 42	INCHEON	-	JUAN	.794E+01
14	N 6	SEOUL	-	DONONG	.556E+01
15	N 13	GWANGJU	-	GEUMSEONG	.446E+01
16	N 36	GONGJU	-	JONGCHONRI	.435E+01
17	N 1	MOGPO	-	MUAN	.432E+01
18	N 24	ULSAN	-	GEOHWARI	.399E+01
19	N 1	GEUMSEONG	-	MUAN	.383E+01
20	N 7	DONGRAE			.324E+01
21	N 7	DONGRAE	-	GEOHWARI	.244E+01
22	N 1	JOCHIWON	-	JONGCHONRI	.235E+01
23	N 3	SEOUL	-	EUJEONGBU	.222E+01
24	N 36	JOCHIWON	-	DONGJEONGRI	.218E+01
25	N 36	CHEONGJU	-	DONGJEONGRI	.217E+01
26	N 5	ANDONG	-	EUSEONG	.185E+01
26	N 5	EUSEONG	-	CHEONPYEONGDONG	.185E+01
28	N 5	DAEGU	-	CHEONPYEONGDONG	.184E+01
29	E 1	PYEONGTAEG	-	CHEONAN	.182E+01
30	E 1	OSAN	-	PYEONGTAEG	.177E+01
31	E 6	MASAN	-	JUNGRI	.174E+01
31	N 3	CHUNGJU	-	JUDEOG	.174E+01
33	E 6	GIMHAE	-	DEOGSAN	.173E+01
34	E 6	MASAN	-	DEOGSAN	.170E+01
34	E 1	SINGAL	-	OSAN	.170E+01
36	E 6	GIMHAE	-	DONGRAE	.169E+01
37	N 28	YEONGCHEON	-	ANGANG	.166E+01
38	N 35	YEONGCHEON	-	IMPO	.158E+01
39	N 28	POHANG	-	ANGANG	.155E+01
40	N 35	IMPO			.154E+01
41	E 1	CHEONAN	-	DONGJEONGRI	.153E+01
42	E 6	JINJU	-	JUNGRI	.147E+01
43	N 35	GYEONGJU	-	ANGANG	.145E+01
44	N 4	GYEONGJU			.140E+01
45	E 3	YEONMUDAE	-	JEONJU	.137E+01
46	N 1	DAEJEON	-	JONGCHONRI	.130E+01
47	N 1	NONSAN	-	YEONMUDAE	.128E+01
48	N 46	CHUNCHEON	-	CHEONGPYEONG	.125E+01
49	L	SEOUL	-	SEONGNAM	.124E+01
50	N 46	CHEONGPYEONG	-	DAESEONGRI	.120E+01
51	E 1	DAEGU	-	YOAEGWAN	.115E+01
52	E 1	GUMI	-	YOAEGWAN	.115E+01
53	E 1	DONGJEONGRI	-	HOEDEOG	.115E+01
54	E 1	DAEJEON	-	HOEDEOG	.110E+01
55	N 23	GONGJU	-	NONSAN	.101E+01

* N = National Highway: E = Expressway: L = Local Road
This highway system classification, established by the Ministry of Construction (MOC), identifies the highway system in Korea. E1, e.g., represents the Expressway Number 1 which connects Seoul to Busan.

Table 6.6: Priority Links for Improvements Based on National Growth Policy: Alternative I

6.7 Priorities for Road Improvements with an Alternative National Growth Policy

In December 1983, the Government of Korea announced its revised fifth Five-Year Development Plan (FYDP), two years after the original fifth FYDP was announced, as was described before. The major implication of the revised fifth FYDP on the distribution of socio-economic activities is the plan's goal for achieving balanced national development. By taking this stance, the government is shifting from a growth pole development policy to a balanced regional growth policy. The programs designed for achieving this goal can be seen in almost all sectors, including education, medical care, and telecommunication, but particularly infrastructure investments and rural developments.

Among the many changes that the balanced regional growth policy may cause, one can reasonably expect that interzonal interaction for both freight and passengers will increase. This level of interaction can be quantified by equation (6.7), as described before. Wilson (1970)[259] termed this measure "entropy" and Erlander (1977)[85] labeled it "the level of dispersion," or "crosshauling."

Acknowledging that there is a one-to-one relationship between investment policy toward balanced regional growth and the level of interzonal economic interactions, a reasonable policy question would then be, "Which transportation facilities should receive higher priority for road improvement if the government wishes to increase the interzonal flow movement level by ten percent?" The question implies that an increase in interzonal interaction by ten percent is a means of achieving the goal of balanced regional growth. Table 6.7 shows alternative priorities for road improvements if the levels of S_r are increased by a few percentage points.

$$S_{freight} = -0.188 \times 10^6 (8.7\% \text{ increase})$$

$$S_{passenger} = -0.726 \times 10^6 (14.3\% \text{ increase}).$$

The highlights of the two different priorities with alternative national growth policies are shown in Figure 6.1, as are links with higher priorities. The notable differences indicate that priorities for road improvements are highly sensitive to national growth policies.

Priority	Highway No.*	Link		Level of Disutility
1	E 9	BUSAN -	GIMHAE	.115E+03
2	E 9	GIMHAE		.806E+02
3	N 37	YANGPYEONG -	YEOJU	.565E+02
4	N 43	GWANGJU (GYEONGGI)-	GWANGDONGRI	.398E+02
5	N 43	SEOUL -	GWANGDONGRI	.387E+02
6	N 1	SUWON -	OSAN	.286E+02
7	N 6	YANGPYEONG		.277E+02
8	N 3	GWANGJU (GYEONGGI)-	ICHEON	.255E+02
9	N 4	GYEONGJU		.157E+02
10	N 1	PYEONGTAEG -	CHEONAN	.127E+02
11	N 1	CHEONAN -	JEONEU	.125E+02
12	N 4	GYEONGJU -	EULDONG	.103E+02
13	N 3	JANGHOWON -	JUDEOG	.960E+01
14	N 3	SEOUL		.832E+01
15	N 3	ICHEON -	JANGHOWON	.773E+01
16	E 6	MASAN -	JUNGRI	.727E+01
17	N 42	ICHEON -	YEOJU	.708E+01
18	N 36	CHEONGJU -	JUDEOG	.652E+01
19	N 42	INCHEON -	SIHEUNG	.533E+01
20	N 35	GYEONGJU -	ANGANG	.522E+01
21	E 1	OSAN -	PYEONGTAEG	.512E+01
22	E 1	DONGJEONGRI-	HOEDEOG	.488E+01
23	N 1	ANYANG -	GUNPO	.465E+01
23	N 1	GUNPO -	SUWON	.465E+01
25	N 1	JOCHIWON -	JEONEU	.414E+01
26	E 1	DAEJEON -	HOEDEOG	.411E+01
27	E 6	GIMHAE -	DEOGSAN	.408E+01
28	E 1	DAEJEON -	OGCHEON	.397E+01
28	N 3	ICHEON		.397E+01
30	N 23	GONGJU -	JEONEU	.391E+01
31	E 1	OGCHEON -	GUCHONRI	.383E+01
31	E 1	HOENGGAN -	GUCHONRI	.383E+01
31	E 1	GIMCHEON -	HOENGGAN	.383E+01
34	L	ANYANG -	SIHEUNG	.378E+01
35	E 7	DAEGU -	DALSEONG	.362E+01
35	E 7	DALSEONG -	HYEONPUNG	.362E+01
35	E 7	HYEONPUNG -	CHANGNYEONG	.362E+01
35	E 7	CHANGNYEONG-	NAMJI	.362E+01
35	E 7	NAMJI -	JUNGRI	.362E+01
40	N 1	SEOUL -	ANYANG	.354E+01
41	E 3	YEONMUDAE -	JEONJU	.350E+01
42	N 42	SUWON -	BANWOL	.347E+01
43	N 3	CHUNGJU -	JUDEOG	.341E+01
44	N 23	MUAN -	YEONGGWANG	.333E+01
45	E 1	GUMI -	YOAEGWAN	.324E+01
45	E 1	DAEGU -	YOAEGWAN	.324E+01
47	E 1	GIMCHEON -	GUMI	.309E+01
48	N 42	SIHEUNG -	BANWOL	.306E+01
49	N 1	JOCHIWON -	JONGCHONRI	.289E+01
50	E 3	SEODAEJEON -	YEONMUDAE	.286E+01
51	N 32	DAEJEON -	GONGJU	.284E+01
52	E 1	SINGAL -	OSAN	.275E+01
53	E 6	MASAN -	DEOGSAN	.270E+01
54	N 36	CHEONGJU -	DONGJEONGRI	.267E+01
55	E 1	PYEONGTAEG -	CHEONAN	.264E+01

* N = National Highway: E = Expressway: L = Local Road
This highway system classification, established by the Ministry of Construction (MOC), identifies the highway system in Korea. E1, e.g., represents the Expressway Number 1 which connects Seoul to Busan.

Table 6.7: Priority Links for Improvements Based on National Growth Policy: Alternative II

6.8 Future Extensions

The importance of developing a proper procedure for setting investment priorities also stems from the fact that investment priorities in developing countries often have to be revised annually. This is due to changes in external factors such as the price of raw materials, balance of payment problems, or natural disasters. In 1983, for example, one year after the adoption of the fifth FYDP, the Economic Planning Board of Korea announced the revised fifth FYDP in response to changes in the external and domestic economic climate. Three important external factors were cited earlier as reasons for the revision: the current decline in raw material prices, especially for crude oil, the unforeseen delay in the world economic recovery, and an improvement in the national balance of payments (Suh 1983)[240].

A possible chain of effects resulting form the external and domestic economic climate may be as follows: changes in external-internal indicators$\rightarrow$ changes in GNP/export targets$\rightarrow$ changes in budgets$\rightarrow$ changes in infrastructure investment$\rightarrow$ changes in priority. A model for formulating road investment priorities should be sensitive to changes in these variables. The model should also explicitly represent interaction between public and private sectors, since both sectors are affected by changes in external and domestic economic climates.

The mathematical formulation of the foregoing analysis can be made as follows:

$$P6.2) \quad \min \sum_a \int_0^{f_a} m_a(x)dx + \sum_i \sum_r p_r^i X_r^i + \sum_{i\varepsilon e} \sum_r d_r^i (E_r^i + I_r^i) \quad (6.24)$$

$$\text{s.t.} \quad \sum_{i\varepsilon e} E_r^i \geq E_r, \sum_{i\varepsilon e} I_r^i \geq I_r \qquad \forall r \quad (6.25)$$

$$I_r^i + \sum_j X_r^{ij} + X_r^i = \sum_j X_r^{ij} + \sum_q a_{rq} X_q^i + E_r^i \qquad \forall r \text{ and } i. \quad (6.26)$$

Thus, when a nation is concerned with the impacts on the transportation system requirements of plan revisions that may become necessary due to external and internal changes in economic activities, minimization of equation (6.24) subject to equation (6.25), (6.26) and (6.5) to (6.8) will result in identifying bottlenecks between pairs of zones. In this case, the assumption is that all transportation modes that are available between the i and j pair are combined into one "type" of transportation mode. The results will still be very useful for planners and analysts in developing countries because the pairwise bottlenecks and investment priorities have been identified as the results of interaction between (1) public sectors that set export (E_r) and import (I_r) targets for the nation and (2) private sectors that choose minimum costs of production ($p_r^i X_r^i$) and shipments (X_r^{ijp}).

6.9 Applications to Developing Countries: General Descriptions

With the alarming rate of increase in costs that are necessary to maintain roads in most developing countries (World Bank 1983[13], p. 45), the model may be useful in identifying priorities for road investments in each of the following three cases:

1. If the country has freight origin and destination shipment data on roads (X_r^{ij}), then the minimization of equation (6.9) subject to equations (6.6) through (6.8), (6.21), and (6.22) will result in the identification of bottlenecks, as demonstrated in the empirical analysis. Bottlenecks thus found can be used in two ways. First, the magnitude of bottlenecks can be directly interpreted as priorities. Second, the bottlenecks identified can be used for generating a *new* alternative network. Any known transportation planning models (static or dynamic) assume that alternative networks are given. The alternative network thus identified by the magnitude of bottlenecks is the result of the system equilibrium principle with explicit representation of congestion on each link.

2. If a country is concerned with a plan for the future, minimization of equation (6.2) subject to equations (6.3) through (6.8) will result in identifying bottlenecks in the existing network when future amounts of shipments are given (with the marginal cost function in the first term in equation (6.2)). The amounts of future shipment can be handled in two ways. First, if amounts to be shipped on the road are known (i.e., mode choice analysis is performed a priori), then amounts by trucks (say, $\sum_j X_{rh}^{ij}$ and by destination ($\sum_j X_{rh}^{ij}$) can be entered into equation (6.4). The implicit assumption here is that the variables $\sum_j X_r^{ij}$ and $\sum_j X_r^{ij}$ in equation (6.4) include all shipment by other modes.

   Second, if a mode choice analysis is not done a priori, the minimization of equation (6.2) subject to equations (6.3) through (6.8) will then result in identifying bottlenecks between pairs of zones, provided that all transportation modes are combined into one "type" of mode. The results will still be very useful for planners and analysts in developing countries. The pairwise bottlenecks are identified as the results of interaction between public sectors, who set export (E_r) and import (I_r) targets for the nation, and private sectors, represented by the agent who chooses minimum costs of production ($p_r^i X_r^i$) and shipments (X_r^{ijp}).

3. Mode choice functions can also be explicitly included in the model, as shown by Kim (1985)[141] in an urban systems context.

6.10 Concluding Remarks

A model for formulating investment priorities for road improvements was developed and tested in this chapter. Interaction between public and private sectors is explicitly represented in the model. The model is also capable of handling changes in external and internal economic indicators and their consequent effects on transport requirements.

The model was applied to Korea using its multiregional-multicommodity flow data, and it was demonstrated that the model can shed light on the issue of identifying investment priorities for road improvements.

The model, however, also has certain limitations. The foremost is that it is static in that it fails to represent the dynamic impacts of improvements of one road on other roads. It also fails to consider the best system, where the "best" is defined in terms of a national welfare function. This problem may be formulated as a bilevel programming problem, as shown in chapters that follows, by combining a welfare maximization problem and a network equilibrium problem.

Nevertheless, it was demonstrated that the model was capable of suggesting priorities for road improvements as the result of interaction between public sectors.

A model for formulating investment priorities for highway improvements under alternative scenarios was also implemented using a detailed transportation network system in Korea. The model resulted in identification of investment priority links with alternative national growth policies. The alternative national growth policy was assumed to be represented by the dispersion measure in a manner that Wilson (1970)[259] and Erlander (1977)[85] discussed.

Part IV

BILEVEL PROGRAMMING MODELS AND APPLICATIONS

Chapter 7

Bilevel Programming

7.1 Introduction

Multilevel decision making is an important aspect of the planning process for planning problems that require synthesis of several interacting individual agencies or subunits.[1] Decision-making agencies may be arranged within a hierarchical administrative structure that has independent or conflicting objectives.

An important feature of multilevel programming problems[2] is their application to multilevel decision making. Multilevel programming allows for a hierarchical structure in which a decision maker at one level of a hierarchy may have an objective function; and decision spaces are determined, in part, by other levels. In addition, decision makers have control instruments that may allow them to influence policies at other levels, thereby improving their own objective functions. These instruments may include the allocation and use of resources at lower levels and the benefits conferred upon other levels. All decision makers optimize their objective functions without considering the effects of their actions on any other decision makers. An additional important feature of multilevel programming problems is that control of the decision variables is partitioned among the decision makers. In traditional single-level programming problems, one decision maker is assumed to have control over all the variables. The usual process in mul-

[1]This chapter is written with Sunduck Suh and draws heavily upon his doctoral dissertation, *Implementation and Evaluation of Nonlinear Bilevel Programming Model of Equilibrium Network Design Problem*, University of Illinois at Urbana-Champaign, Urbana, IL. January, 1989.

[2]A bilevel programming problem is a special case when only two levels are involved.

tilevel public sector decision making, however, is not simultaneous, but sequential, down the administrative hierarchical decision structure (Bialas and Karwan 1984)[34]. For example, in many countries the central government first allocates resources among different provinces, and then each provincial government decides its own action based on the resources allocated by the central government. In a mixed economic system, the public sector supplies transportation facilities and the private sector adjusts itself to these given transportation facilities. Decision makers, with the goal of optimizing their objective functions, set values for the decision variables under their control after people at higher levels set their own values on the same variables.

To date, most applications of multilevel programming have been focused on modeling economic problems. Recently, new applications have started to emerge in such fields as development planning, agricultural policy problems, planning process problems, and network design problems. The proliferation of multilevel programming in several fields suggests that multilevel programming approaches have advantages over single-level programming. The advantages include the ability to explicitly model the sequential decision-making process and interactions between different levels of decision processes or systems.

The ability of multilevel programming to explicitly model the segmented decision making and the interactions between the parties involved makes multilevel programming an ideal tool for modeling the transportation investment decision process, which involves the private-public sector's interactive and/or joint decisions.

The following section more specifically defines the bilevel programming method and describes its pertinent properties. A discussion of the applications of bilevel programming then follows, preceded by a description about the development of algorithms for bilevel programs.

7.2 Definition and Properties of Bilevel Programming

7.2.1 Definition

In general, a bilevel programming problem (Kolstad 1985[146]; and Boyce and Kim 1987[41]) is defined as follows:

$$P7.1) \quad : \quad \mathbf{U7.1)} \quad \min_{x} \quad F(x, y) \tag{7.1}$$

$$\begin{aligned} \text{s.t.} \quad & G(x,y) \leq 0 && (7.2) \\ \text{where} \quad & y(x) \quad \text{is defined by} \\ \textbf{L7.1)} \quad & y(x); \min_{y} \quad f(x,y) && (7.3) \\ \text{s.t} \quad & g(x,y) \leq 0. && (7.4) \end{aligned}$$

U7.1) is defined as an upper level[3] problem and L7.1) a lower level[4] problem. The decision maker at the upper level influences the lower-level decision maker by setting x, thus restricting the feasible constraints set for the lower-level decision maker. The upper-level decision maker also interacts with the lower-level decision maker via the objective function of the lower-level decision maker. Also note that the decision variable of the lower-level problem is expressed as a function of the decision variable of the upper-level $(y(x))$.

7.2.2 Feasibility and Optimality

This subsection is based on Bard (1983[16], 1984[21]). The decision maker at the upper level has control over the set $X \in R^{n^1}$ of decision variables while the decision maker at the lower level has control over the set $Y \in R^{n^2}$ of decision variables. It is assumed that $X \cap Y = \emptyset$. With bilevel programming, the decision maker at the upper level acts first by choosing x. This choice may affect the set of feasible strategies available to the lower-level decision maker. For fixed value x, the problem faced by the lower-level decision maker becomes

$$\min \quad (f(x,y) : y|x) \qquad (7.5)$$

s.t.

$$x, y \in S = \{(x,y) : G(x,y) \leq 0, g(x,y) \leq 0\}. \qquad (7.6)$$

The solution set of the lower-level decision maker is referred to as the *rational reaction set* and is defined as follows:

Definition 1 (Bard 1983.) *The rational reaction set of f over S as defined in equation 7.6 is given by*

$$W_f(S) \equiv \{x \in X, y \in Y : (x,y) \in S, f(x,y) = \min(f(x,y) : y|x)\} \qquad (7.7)$$

[3] Also called as a outer, out-side, level one, leader, or policy problem (see Candler and Townsley)[56].

[4] Also called as a inner, in-side , level two, follower, or behavioral problem (see Candler and Townsley)[56].

Furthermore, for each $\bar{x}$, the rational reaction is the mapping

$$\bar{y} = \Phi_f(\bar{x}) \tag{7.8}$$

if there existed a $\bar{y}$ that uniquely minimizes $f(\bar{x}, y)$ for all $(\bar{x}, y) \in S$.

The problem faced by the decision maker at the upper level with control over both variables becomes

$$\min(F(x,y) : (x,y) \in W_f(S)) \tag{7.9}$$

or

$$\begin{aligned} \min F(x,y) \quad &: \quad (x,y) \in \bar{S} && (7.10) \\ \bar{S} \quad &= \quad \{(x,y) : x \in P, y \in Y(x)\} && (7.11) \\ \text{and } P \quad &= \quad \{x \in X : G(x,y) \leq 0, \exists y \in Y, g(\bar{x},y) \leq 0.\} && (7.12) \end{aligned}$$

Definition 2 (Bard 1983.) *A point $\bar{y}$ is said to be optimal with respect to a point $\bar{x}$ if $\bar{y} = \Phi_f(\bar{x})$; such a pair $(\bar{x}, \bar{y})$ is said to be* feasible.

Definition 3 (Bard 1983.) *A point (x^*, y^*) is said to be an* optimal solution *to the bilevel programming if*

1. *(x^*, y^*) is feasible, and*

2. *for all feasible pairs $(\bar{x}, \bar{y}) \in S$,*

$$F(x^*, y^*) \leq F(\bar{x}, \bar{y}) \tag{7.13}$$

(S is usually assumed to be nonempty and compact.)

By substituting the lower-level problem with its Karush-Kuhn-Tucker condition, the bilevel problem can be reformulated into a single-level problem. The first-order necessary condition of the bilevel programming problem is as follows:

Theorem 1 (Bard 1983.) *Let F, G, f, and g be at once continuously differentiable and assume that a constraint qualification holds for equation (7.3) and equation (7.4) with x fixed as x^*. Then a necessary condition that (x^*, y^*) solves* **P7.1** *is that there exist a $w^* \in R^m$ such that (x^*, y^*, w^*) is* *feasible to the following problem:*

$$\min_{x,y,w} \quad F(x,y) \tag{7.14}$$

$$s.t. \quad G(x,y) \geq 0 \tag{7.15}$$

$$\nabla_y f(x,y) + w\nabla_y g(x,y) = 0 \tag{7.16}$$

$$w * g(x,y) = 0 \tag{7.17}$$

$$g(x,y) \geq 0 \tag{7.18}$$

$$w \geq 0 \tag{7.19}$$

where w is the m-dimensional vector of Karush-Kuhn-Tucker multiplier associated with the inside problem and ∇ is the gradient operator.

Note that equation (7.16) to equation (7.19) implicitly represent the rational reaction set $W_f(S)$.

7.2.3 Computational Complexity

Bilevel programming problems in general defy easy calculation procedures. One cause for this computational complexity is explained by the NP-hardness of the bilevel programming problem. NP-hardness of bilevel programming problem is demonstrated by Ben-Ayed (1988)[27] based on the work of Jeroslow (1985)[128]. NP-hardness implies that no polynomial algorithm exists to solve the bilevel programming problem.

The nonconvexity of bilevel programming presents another problem. Nonconvexity might result even though both lower- and upper-level problems are convex problems. This nonconvexity of the linear bilevel problem can be best illustrated by the example from Bialas and Karwan (1984)[34]. The example shows that the feasible region of the linear bilevel programming need not be convex, because the feasible region is composed of a set of faces of feasible polyhedron S.

For a nonlinear problem, it is a nonconvex problem if the condition of equation (7.17) shows the nonconvexity. Nonconvexity implies that even if the solution to the problem can be identified, the solution may be only a local solution rather than a global one.

An inherent drawback of the linear bilevel programming problem is that the lower-level problem can have multiple optima. Note that in developing the rational reaction set, given the upper-level decision variable (equation (7.8)), a unique optimum for the lower-level problem is required. Bialas and Karwan (1978)[33] suggested one scheme to overcome this problem. The problems considered in this book do not have multiple optima since they are all nonlinear programming problems.

On the bright side, the bilevel programming problem has some interesting characteristics that are inherent in game theory. Relying on game theory, one can possibly bound the objective function value of the bilevel programming problem. This bound of the objective function can be utilized in developing heuristic algorithms.

7.2.4 Bounding of Objective Function Values

It is well known from game theory that if one player has control over all the decision variables, the player can better optimize his or her own objective function than in the case where control is split between two players (Simaan and Cruz 1973)[227]. The bilevel programming problem P7.1) can be reformulated into three different problems depending who has control over what variables. The following problem P7.2) results if all the decision variables are controlled by the upper-level decision maker:

$$P7.2) \qquad \min_{x,y} \quad F(x,y) \tag{7.20}$$

$$G(x,y) \leq 0 \tag{7.21}$$

$$g(x,y) \leq 0. \tag{7.22}$$

The following formulation represents the case where both the lower-level decision maker and upper-level decision maker have control over their own variables and do not know the other's reaction to their variable choices. In this situation, both decision makers' behavior is myopic.

$$P7.3) \qquad \min_{x} \quad F(x,y) \tag{7.23}$$

$$G(x,y) \leq 0 \tag{7.24}$$

where

$$\min_{y} \quad f(x,y) \tag{7.25}$$

$$g(x,y) \leq 0. \tag{7.26}$$

Notice that in P7.3), y is no longer a function of x.

Specifically, P7.1) results when the decision variables are partitioned into lower-and upper-level decision makers and the leader knows the reaction of the follower. With the solution of three formulations, (x_1^*, y_1^*), (x_2^*, y_2^*) and

(x_3^*, y_3^*) for P7.1), P7.2) and P7.3), respectively, the objective function values $F(x_1^*, y_1^*)$ are bounded as follows:

$$F(x_2^*, y_2^*) \leq F(x_1^*, y_1^*) \leq F(x_3^*, y_3^*). \tag{7.27}$$

For most of the cases, the solution for the single-level problem of P7.2) or P7.3) is much easier to solve than P7.1). Therefore, instead of solving P7.1) directly, one can solve P7.2) or P7.3) and try to reduce the gap between $F(x_1^*, y_1^*)$ and $F(x_2^*, y_2^*)$, or $F(x_3^*, y_3^*)$. Heuristic algorithms (see Shaw 1980 [223] and Ben-Ayed 1988 [27]) approach the problem this way.

Another bounding scheme is to utilize the concept of Cournot-Nash equilibrium. The bilevel programming problem is a special case of Stakelberg equilibrium. The upper-level player is assumed to know the reaction of the lower-level player to the decision of the upper-level player. In the Cournot-Nash game, on the other hand, both players do not know the reaction of the other players to the their decision.

With the solution of (x_N^*, y_N^*) and (x_1^*, y_1^*) for the Cournot-Nash equilibrium and bilevel programming P7.1) (Stakelberg equilibrium) respectively, objective functions $F(x_1^*, y_1^*)$ and $F(x_N^*, y_N^*)$ have the following relationship (Moulin 1982)[188]:

$$F(x_1^*, y_1^*) \leq F(x_N^*, y_N^*). \tag{7.28}$$

In fact, the problem P7.3) models the Cournot-Nash equilibrium. The application of this bounding scheme to the solution of bilevel network design problems has been studied by Marcotte (1983)[173] and Harker and Friesz (1984)[115].

7.2.5 Relation with Other Techniques

In this subsection, multilevel programming is compared with decomposition technique, hierarchical planning methodology, goal and multi-objective programming, game theory, control theory, and equilibrium programming. Some of the discussion in this subsection is based on Nwosu (1983)[197].

Decomposition. The term "multilevel programming" has also been used to describe the result of the decomposition technique (Dantzig-Wolfe 1960[76]; Balas 1966[12]; Cooper et al. 1964[67]; Geoffrion 1970[107]; Lasdon 1970[157]; and Kornai 1973[148]) on large-scale systems. The decomposition technique provides for incorporation of a single decision maker who has a single objective function and controls all the decision variables (Dirickx and Jennergren 1979)[81]. The solution process can be interpreted to analyze the behavior of a multilevel organization; however, it is only solving an optimization problem with a single objective function over a fixed feasible region. The assumption is that either the single objective decomposes into the objectives of the subunits or that the central planner is willing to accept the aggregated objectives of the subunits as his or her own. The basic idea of multilevel decomposition technique envisions an iterative process in which the decision maker repeatedly sends out data and receives feedback to upgrade the current system solution.

Hierarchical Planning Method. Hierarchical planning methodology is a term also proposed by Hax and Meal (1975)[117] from the field of production control. In their work, the overall problems have been partitioned into a number of subproblems according to classifications of the product structure. In this formulation, however, the direction of the information flow is only downward, while multilevel programming provides for a two-way flow of information.

Goal and Multi-objective Programming. Goal and multiobjective programming also deal with problems having multiple objective functions. Both goal programming and multi-objective programming, however, assume the objective functions to be those of a single decision maker or those of a collection of harmonious decision makers. For instance, in goal programming, the objective functions are ordered in a preference hierarchy and optimized in sequence (Ignizio 1976[123]; Cohon 1978[66]). The solution techniques for multi-objective programming problems seek Pareto optimal solutions (Keeney and Raiffa 1976)[131]. This is equivalent to optimizing some convex combination of the objective functions (Isermann 1974)[127]. Thus, the solution is identified by reaching a compromise among the competing objectives. This is in contrast to multilevel programming, where these objective functions are those of independent decision makers at various levels of an organization.

Game Theory. In game theory, the usual assumptions are that all players move simultaneously and that their strategy spaces are independent and disjointed. It is possible, however, to prescribe a game requiring a set of sequential moves, provided each player employs a Stackelberg strategy (Simaan and Cruz 1973[227]). Even though the decision variables are partitioned among the players (decision makers), current game theory does not consider the case where the payoffs and strategy spaces for players can be altered by independent actions of the other players. An appropriate extension of Stackelberg games would require the payoff function of the lower-level to have discontinuities dependent on the decisions of the other players.

Control Theory. In principle, control theory is concerned with the evolution in time of a stochastic dynamic system where the form of the state variable, the function determining the evolution, is known. But multilevel programming is a static nonstochastic system, not a stochastic dynamic system. In multilevel programming, therefore, the determination of the functional form specifying the system equilibrium is central to the problem.

Equilibrium Programming. The closest formulation to multilevel programming is equilibrium programming. In equilibrium programming, however, all decisions are made simultaneously, and therefore hierarchial decisions cannot be modeled (Zangwill and Garcia 1981)[265]. Bard (1983)[17] showed the relationship between multilevel programming and equilibrium programming.

7.3 Applications of Bilevel Programming

Bilevel programming models have a fairly long list of applications, especially in the economic discipline. For instance, it has been applied to the natural resource management problem (Schenk 1980[219]; Kolstad 1986[145]), project selection (Cassidy et al. 1971)[57], strategic planning for water conservation and distribution policies (Bisschoop et al. 1982)[35], strategic planning for the agriculture sector (Candler et al. 1981)[55], and regional development problems (Seo and Sakawa 1980)[222]. See Kolstad (1985)[146] for a more detailed annotated bibliography on bilevel programming applications up to 1982 and Kim and Suh (1988)[142] on more recent works.

Bilevel programming applications of the network design problem can be classified into two categories. The first is the formulation of the problem in a linear bilevel network design problem that includes the work of LeBlanc and Boyce (1986)[159], Ben-Ayed, Boyce, and Blair (1988)[28], and Ben-Ayed (1988)[27]. The second category is the formulation of the problem into a nonlinear problem that includes the work of Marcotte (1986)[172], Marcotte and

Marquis (188)[175], Kim and Suh (1988)[142], Suh and Kim (1988)[238], and LeBlanc and Boyce (1986)[159]. Marcotte (1986)[172] proposed a nonlinear bilevel network design model where the lower-level problem has been substituted with an equivalent variational inequality problem. Later, Kim and Suh (1988)[142] proposed a nonlinear bilevel network design problem that uses a combined distribution and assignment problem with an entropy constraint as the lower level-problem. In addition to a linear bilevel formulation, LeBlanc and Boyce (1986)[159] proposed a nonlinear bilevel network design problem that utilizes a user-equilibrium route choice problem as the lower-level problem. Marcotte (1986)[172] and Marcotte and Marquis (1988)[175] reported computational results using a small network problem with five heuristic algorithms. Suh and Kim (1988)[238] report computational results that were obtained by a descent-type algorithm. They examined the performance of the proposed algorithm based on the computational results for several examples, as described in the section that follows. The algorithm is a simplified real-life network that has seven origin-destination zones, fifty-seven nodes, and 172 links.

7.4 Solution Algorithms for Bilevel Programming

Algorithmic approaches developed for solving bilevel programming problems can be classified into four types. These are: extreme point search, Karush-Kuhn-Tucker method, descent method, and the direct search method, such as the Hooke-Jeeves method used in Abdulaal and LeBlanc (1979)[2]. The following section briefly discusses algorithms in each category.

Extreme Point Search Method. This approach has been used for solving linear bilevel problems and is based on the claims that any solution to the linear bilevel programming problems occurs at an extreme point of the constraints set for the lower-level problem (Bialas and Karwan 1981[34]; Candler and Townsley 1982[56]; Nwosu 1983[197]).

Assuming that the constraint space is bounded and that a unique solution exists for the lower-level problem for any feasible upper-level variables, several algorithms have been developed for solving bilevel linear resource control problems. Local optimal solution procedures have been proposed by Bialas and Karwan (1978[33], 1984[34]) and by Candler and Townsley (1978)[56]. Candler and Townsley (1982)[56] have also presented a global optimal solution procedure. Their procedure includes an implicit exploration of all combinations of vectors that could form a dual feasible basis for the upper-level problem. As studied by Nwosu (1983)[197], however, this approach fails to guarantee the global optimum solution.

Bialas and Karwan (1981)[34] have shown that the solution of the problem must occur at an extreme point of the constraint space. Based on this observation, they proposed the "kth-best" algorithm, which finds the global optimal solution by an explicit, though partial, enumeration of the extreme points of the constraint space (Wen 1980[255]; Wen 1981[256]; Bialas and Karwan 1984[34]).

Karush-Kuhn-Tucker Approach. Another solution approach is for the bilevel linear resource control problem. This approach involves replacing the lower-level problem by its Karush-Kuhn-Tucker conditions (Bard and Falk 1982[20]; Fortuny and McCarl 1981[98]; Shaw 1980[223]; Bialas and Karwan 1984[34]). Bard and Falk (1982)[20] solved the bilevel linear problem using a nonconvex programming algorithm based on branch and bound techniques. The feasible region is enclosed within a linear polyhedron, which is then partitioned. A global solution is obtained from a piece-wise linear approximation of the nonlinear problem. However, as reported by Bard and Falk (1982)[20], they had only limited computational experiences.

Fortuny and McCarl (1981)[98] included price control aspects in the lower-level problem and enforced the complementary conditions by transforming the Karush-Kuhn-Tucker conditions into a much larger mixed integer programming problem. They presented empirical results of fertilizer dealers. Using this formulation, Schenk (1980)[219] solved a bilevel pollution control model and showed that, because of the large size of the augmented problem, only small problems could be reasonably solved.

Shaw (1980)[223] developed the Parametric Complementary Pivot (PCP) algorithm. The PCP algorithm can be viewed as an implicit enumeration of the lower-level optimal bases. All of the computations, however, may be performed within the framework of a single tableau, which requires less memory space than the kth-best algorithm. Nevertheless, Blair found a counterexample to the PCP algorithm.

Algorithms developed by Bard and Moore (1987)[22] and Moore and Bard (1987)[183] also fall into this category. The algorithm proposed by Bard and Moore (1987)[22] is based on the branch and bound approach to enforcing the underlying complementary slackness condition of the Karush-Kuhn-Tucker conditions. The Moore and Bard algorithm (1987)[183] solved for the zero-one bilevel programming problem based on the implicit enumeration approach. Even though computation results using randomly generated problems seem to be promising in terms of computer CPU time, the applicability of these two algorithms to real-world problems has still not been proven. In addition, the computation time is expected to increase exponentially as the number of variables increases, because their approach is basically an implicit enumeration approach.

Bialas and Chew (1981)[32] gave an efficient algorithm for the K-level continuous knapsack problem, where each decision maker controls exactly one variable. Although this is an extremely simple model that can easily be solved by dynamic programming, the main thrust of their papers was to study the formation of a coalition among subsets of the decision makers to obtain a Pareto optimal solution. Wen (1981)[256] proposed a hybrid algorithm consisting of the kth-best and the PCP algorithm. He attempted to solve the three-level resource control programming problem. His "global" optimum solution for the problem, however, was proven by Nwosu(1983)[197] to be a suboptimal solution.

Narula and Nwosu (1982[191], 1983[192]) attempted to solve the bilevel and K-level multilevel problems. They explicitly generated the constraint sets for the upper-level decision maker and then employed a branch and bound technique to solve the problem. The proposed branch and bound procedure, however, did not guarantee global optimal solutions. Bard (1982[18], 1983[20], 1985[19]) developed the Grid Search Algorithm (GSA), which had been regarded as the best algorithm until Ben-Ayed, Boyce, and Blair (1988)[28] found a counter-example that showed the GSA fails in certain circumstances (see also Marcotte 1988)[174].

Another recent development is Pre-emptive Hierarchical Programming (PHP) developed by Nwosu (1983)[197] (see also Narula and Nwosu 1985[190]). Again, Blair provided a counter-example.

Descent Method. This approach is based on gradient information, acquired in a variety of ways, from the lower-level problem. The approach has been used mainly for solving nonlinear bilevel problems. This method is an iterative solution technique. It generates a sequence of solution points that can reduce the objective function value, utilizing the gradient information of the lower-level problem with respect to the upper-level decision variables.

An algorithm called the barrier method was developed by Shimizu and Aiyoshi (1981)[226]. Kolstad and Lasdon's (1986)[147] approach for descent method seems more efficient than de Silva's (1978)[79] algorithm in solving a large-scale problem because their approach requires a solution of a smaller problem in derivative calculation. The largest problem size that was reported to have been solved involved 230 variables and thirty constraints.

Direct Search Method. Direct function minimization algorithms, such as the Hooke-Jeeves algorithm utilized by Abdulaal and LeBlanc(1979)[2], belong to this category. These methods rely only on the behavior of the upper-level objective function value in searching for the solution.

The applicability of these available solution algorithms for the bilevel network design problem has been studied by Kim and Suh (1988)[142]. They concluded that none of these available algorithms can be directly applied to the solution of the bilevel network design problem used in this book. Specifically, they noted that some of the algorithms for the linear bilevel programming problem did not work. Furthermore, they also advocated the use of a nonlinear model over a linear model because of the huge problem size of a linear model.

7.5 Concluding Remarks

Bilevel programming is a special kind of multilevel programming. It can explicitly model sequential decision making among different levels of decision making.

In this chapter, bilevel programming problems have been defined, along with feasibility and optimality conditions. Some interesting characteristics of bilevel programming problems, such as nonconvexity, the bounding scheme of the upper-level objective function value, and the relationship with other techniques have been described; and solution algorithms for bilevel programming problems have been summarized with the conclusion that no current available algorithms are directly applicable to the solution of the bilevel network design problem used in this book.

The next chapter actually applies bilevel programming to network design problems. Specifically, it shows how bilevel programming can be used to model an equilibrium network design problem.

Chapter 8

Bilevel Transportation Network Design Models and Solution Algorithms

8.1 Introduction

As described in the previous chapter, there are at least two parties involved in the transportation planning process.[1] The first is the private sector, which adjusts itself to the given level of transportation facilities in an optimal way based on its own behavioral criteria. This behavior represents the demand-side of the process. The other party is the public sector, which supplies the transportation facilities. The public sector provides transportation facilities in such a way that the social welfare function is maximized. The public sector is the supply side.

An assignment process that loads traffic demand between the origin and destination of a network is an example of a demand side transportation modeling activity. The level of service of a given network can be identified under demand by the assignment process. This mathematical model usually takes the form of a multicommodity flow model (Assad 1978)[9]. When user behavior is assumed to follow Wardrop's [252] first principle, the assignment process is called a user-equilibrium assignment (Wardrop 1952)[252]. There

[1] This chapter is written with Sunduck Suh and draws heavily upon his doctoral dissertation, *Implementation and Evaluation of Nonlinear Bilevel Programming Model of Equilibrium Network Design Problem*, University of Illinois at Urbana-Champaign, Urbana, IL. January, 1989.

has been a great deal of research in this area (Boyce 1984)[38]. In addition to the standard route-choice model with fixed demand, some models do not require the demand to be fixed and are referred to as elastic demand models (Gartner 1980)[106], while other models include entropy (Wilson 1970)[259] in the constraints. Still other approaches decide travel demand endogenously in the model (Kim 1983)[135].

The demand models, however, assume that the transportation facilities are fixed. This means that these models do not have features to recommend how many more facilities should be provided to meet the demand given a specified level of congestion. They simply calculate the level of congestion given transportation facilities, regardless of whether the demand is fixed, elastic, or endogenously generated. Therefore, an alternative model is necessary in order to properly access the impact of changes in facilities on the network.

It is possible to use these demand models to forecast the impact of an additional small number of link segments or of a proposed plan. It is not possible, however, to use the demand model to identify the best plan for link addition that involves, say, ten link additions. In theory, such a plan requires 1,024 evaluations of a demand model to identify the best possible combination of ten links involved.[2] Furthermore, the kind of link addition, whether two-lane or four-lane, for example, still has to be decided exogenously. The situation will be even worse if the involved planning variables can take continuous values. The possible combinations become explosive.

On the other hand, the network design model, which is a supply model, can identify the best alternatives in an efficient way compared to what demand models can do. A network design model can identify the optimal link addition/deletion[3], the optimal level of link expansion. The former is called a *discrete* network design problem and the latter a *continuous* network design problem. The assumption of fixed demand can also be relaxed. Therefore, the network design model is appropriate for identifying the best alternatives or combination of alternatives when many policy alternatives are considered for evaluation.

Computational difficulty has been the primary deterrent preventing wide utilization of network design models in actual practice.[4]. Furthermore, the supply of additional links can actually increase the total transportation cost

---

[2] $2^{10} = 1024$.

[3] With proper assumptions about cost, this is also possible to model.

[4] In fact, Lenstra et al. (1976)[163] proved that the discrete network design problem is an NP-complete problem

if the reaction of the user to the additional link is not considered. This well-known phenomenon is called Braess' paradox (see Murchland 1970)[189]. This paradox can be prevented by considering the behavioral assumption of the users in the network design problem. Therefore, it is natural to seek a way to incorporate both private sector (demand) and public sector (supply) and to accommodate the interaction between private and public sector in one network design problem (Boyce 1986[36]; Boyce and Kim 1987[41]; Kim and Suh 1988[142]).

Furthermore, this interaction between public and private sector happens in a hierarchical manner. The public sector provides the facilities and the private sector adjusts to the given facilities. Therefore, the use of bilevel programming, which can explicitly model hierarchical decision making for illustrating the public-private sector interaction, is very adequate. This is the rationale behind the utilization of nonlinear bilevel programming models to formulate network design problems. This new approach will further enhance the applicability of network design problems in transportation planning issues.

The purpose of this chapter is to formulate a network design problem for transportation planning in the framework of the bilevel programming technique. Toward this end, the network equilibrium problem will account for the demand side, the network design problem will model the supply side, and the equilibrium network design problem will represent a combination of the two.

8.2 Network Equilibrium Problem

The network equilibrium problem usually refers to an equilibrium route choice problem in a given network. The solution to an equilibrium route choice problem is to find an equilibrium link flow pattern that satisfies Wardrop's two equilibrium conditions (Wardrop 1952)[252]as was introduced in Chapter 5. These conditions are that

1. all routes that are used between each O/D pair have equal cost; and

2. no driver can reduce his travel cost by switching routes.

This equilibrium flow patten is demonstrated in Figure 8.1 (Gartner 1980)[106]. Two parallel links (a and b) are assumed to connect one O/D pair where user-optimized route choice is obtained at point E, the intersection of the two average link cost functions. Thus, the user-optimizing solution

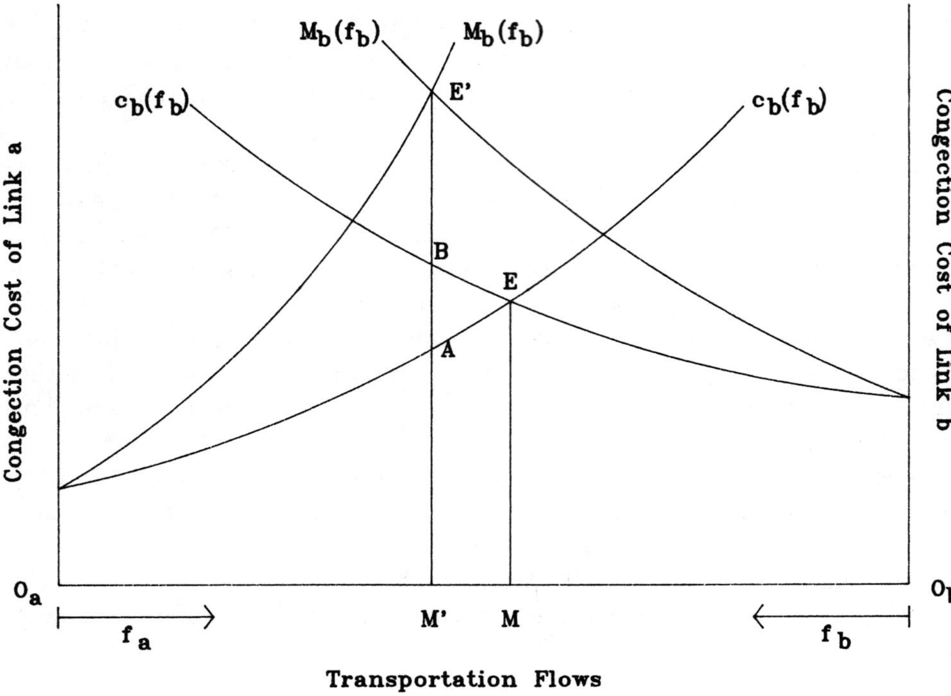

Figure 8.1: User Equilibrium Versus System Equilibrium

for link a (f_a) and link b (f_b) is obtained at point M. The average travel cost on each link is thus EM. Thus user-optimization network equilibrium costs of revery link. A user-optimization network equilibrium does not, in general, achieve the economic allocative efficiency in which the aggregated travel disutility is minimized.

The view of economists is that in order to achieve economic efficiency, every user should pay the full cost for the service obtained. Therefore, the increment between average and marginal cost should be charged as a toll by the operator of the transportation system (Gartner 1980)[106]. An economically efficient choice set, the system-optimization equilibrium, is then achieved at point E', the intersection of the marginal cost curves in Figure 8.1. In this case, the measures of average cost on the two linkis are unequal, as shown in AM' and BM', but the system has been achieved. This new system-optimal flow minimizes the total system cost.

In an optimization problem, the equilibrium flow identifying process was first formulated by Beckmann, McGuire, and Winsten (1956)[24] as follows:

$$P8.1) \quad \min_x \quad \sum_a \int_0^{f_a} c_a(x)dx \tag{8.1}$$

$$\text{s.t.} \quad \sum_p X^{ijp} = T^{ij} \tag{8.2}$$

$$X^{ijp} \geq 0 \tag{8.3}$$

$$f_a = \sum_i \sum_j \sum_p X^{ijp} \delta_a^{ijp} \quad \forall a \tag{8.4}$$

Where $c_a(x)$: average travel cost function on the link a

X^{ijp} : flow on path p from i to j

T^{ij} : flow from i to j

f_a : flow on link a

δ_a^{ijp} : $\begin{cases} 1 & \text{if link } a \text{ is on path } p \text{ from } i \text{ to } j \\ 0 & \text{otherwise.} \end{cases}$

The objective function, equation (8.1), is to minimize the sum of the areas under the average travel cost function. This objective function does not have any economic meaning. The constraints equation (8.2) and equation (8.3) are flow conservation and non-negative constraints.

This P8.1) formulation has many interesting characteristics. Its relationships to Wardrop's conditions (Wardrop 1952)[252] and the condition that guarantees the uniqueness of the solution are well known. For later reference, these two characteristics are briefly summarized here, based on Sheffi (1985)[224].

The Lagrangian of the **P8.1** is

$$\begin{aligned} L(X,u) &= \sum_a \int_0^{f_a} C_a(x)dx + \sum_i \sum_j u^{ij}(T^{ij} - \sum_p X^{ijp}) \\ &\quad - \sum_i \sum_j \sum_p \theta^{ijp} X^{ijp}. \end{aligned} \tag{8.5}$$

To have a solution, the following first-order conditions of P4) should be met at the stationary point of the Lagrangian L:

$$X^{ijk}\frac{\partial L(X,u)}{\partial X^{ijk}} = 0 \tag{8.6}$$

$$\frac{\partial L(X,u)}{\partial X^{ijk}} \geq 0 \tag{8.7}$$

$$\frac{\partial L(X,u)}{\partial u^{ij}} = 0 \tag{8.8}$$

$$X^{ijk} \geq 0. \tag{8.9}$$

The above equations can be reduced to the following first-order conditions:

$$X^{ijk}(C^{ijk} - u^{ij}) = 0 \quad \forall i,j,k \tag{8.10}$$

$$C^{ijk} - u^{ij} \geq 0 \quad \forall i,j,k \tag{8.11}$$

$$\sum_k X^{ijk} = T^{ij} \quad \forall i,j \tag{8.12}$$

$$X^{ijk} \geq 0 \quad \forall i,j,k. \tag{8.13}$$

These conditions correspond to Wardrop's equilibrium conditions. If there is a flow on path k joining i to j ($X^{ijk} > 0$), then the path cost (c^{ijk}) is the same as the equilibrium cost (u^{ij}) from equation (8.10) and equation (8.11). The path cost (c^{ijk}) of the unused path ($X^{ijk} = 0$) cannot be lower than the equilibrium cost (u^{ij}) based on equation (8.10) and equation (8.11).

To show the uniqueness of the solution for the formulation, it is sufficient to show that the objective function is convex. This convexity can be shown with the Hessian matrix of the objective function.

$$Z(X) = \sum_a \int_0^{f_a} c_a(x)dx. \tag{8.14}$$

The gradient of Z(x) is

$$\frac{\partial Z(x)}{\partial x_m} = c_m(x_m) \quad \text{for } m^{\text{th}} \text{ link.} \tag{8.15}$$

Assuming the link cost is dependent only on the flow on the link, the Hessian of Z(x) is

$$\frac{\partial^2 Z(x)}{\partial x_m \partial x_n} = \frac{\partial c_m(x_m)}{\partial x_n} = \begin{cases} \frac{dc_n(x_n)}{dx_n} & \text{for } m = n \\ 0 & \text{otherwise.} \end{cases} \tag{8.16}$$

Therefore, the resulting Hessian matrix is a diagonal matrix with the diagonal element of $\frac{dc_a(x_a)}{dx_a}$.

$$H = \nabla^2 Z(x) = \begin{bmatrix} \frac{dc_1(x_1)}{dx_1} & 0 & \cdots & 0 \\ 0 & \frac{dc_2(x_2)}{dx_2} & \cdots & 0 \\ \vdots & \vdots & \ddots & \vdots \\ 0 & 0 & \cdots & \frac{dc_n(x_n)}{dx_n} \end{bmatrix} \tag{8.17}$$

For this H to be positive definite, each diagonal element should be a positive number (Avriel 1976)[10]. Therefore, the following conditions should be met for the congestion function:

$$\frac{dc_1(x_1)}{dx_1} > 0. \tag{8.18}$$

In other words, the congestion function should be nonnegative and monotonically increasing. This is the condition for **P8.1)** to have a unique solution. This condition has implications for the development of the link congestion function in chapter 6. Unfortunately, however, this convexity does not hold for path flow (see Sheffi 1985[224] for more detail). This non-uniqueness of the path flow solution of the problem has been a deterrent to the development of derivative calculation procedures and subsequently to the application of the descent method to solve the bilevel network design problem.

To obtain a system optimum solution, it is sufficient to substitute equation (8.1) with

$$\min_x \sum_a \int_0^{f_a} M_a(x)dx \tag{8.19}$$

$$\text{where} \quad M_a(x) = \frac{d}{dx}[c_a(x) \cdot x].$$

The area under the total travel cost function is minimized with this objective function. Therefore, the solution actually minimizes the total system cost.

With the advent of a convergent algorithm to the problem, the formulation of **P8.1)** has had profound impacts on the study of equilibrium route-choice problems. Dafermos(1968)[69] proposed an algorithm that requires path enumeration. Leventhal et al. (1973)[165] reported a column generation algorithm. This algorithm does not require the a priori enumeration of all paths but generates the best paths periodically. LeBlanc et.al (1975)[160] solved the problem using the convex combination proposed by Frank-Wolfe (1956)[99]. The advantage of using the Frank-Wolfe algorithm in solving **P8.1)** is that, given a feasible nonoptimal flow, an improved feasible flow direction can be easily found by solving the linear network flow problem, which is equivalent to finding the shortest path in every iteration. Nguyen (1974)[195] proposed the convex-simplex method.

Based on the **P8.1)** formulation, there have been many variations. For instance, Boyce et al. (1981)[63] studied a series of combined problems. Kim (1983)[135] proposed a land use-transportation model based on the equilibrium problem concept. Stochastic equilibrium has been proposed by Daganzo and Sheffi (1977)[75], Sheffi and Daganzo (1980)[225], and Fisk (1980)[95]. For a more through review, see Boyce, LeBlanc, and Chon (1988)[42].

In addition to these variations of nonlinear programming formulation, models have been formulated using other mathematical programming techniques. Aashitiani and Magnanti (1981)[1] proposed a nonlinear complementarity problem formulation for the equilibrium problem:

$$x \cdot F(x,u) = 0 \tag{8.20}$$

$$F(x,u) = \begin{bmatrix} c(x) - u^{ij} \\ \sum X^{ijp} - T^{ij} \end{bmatrix} \geq 0 \tag{8.21}$$

$$x = (x,u) \geq 0. \tag{8.22}$$

Smith (1979)[228], Dafermos (1980)[72], and Fisk and Boyce (1983)[94] studied the variational inequality (VI) formulation.

$$c(x^*)(x - x^*) > 0 \tag{8.23}$$

$$\text{for all} \quad x \in \Omega \tag{8.24}$$

$$\Omega = \{x | \Lambda x = T, x > 0\}. \tag{8.25}$$

The possibility of using these variation models in a bilevel programming framework is explored later in the book.

8.3 The Network Design Problem

The network design problem is concerned with the efficient use of limited resources. With respect to strategic, long-term transportation planning, the network design problem solves for the strategic capital investment decisions for highways, airports, and other transportation-related activities. These decisions have impacts on the infrastructure of transportation system networks and, subsequently, the resources required for infrastructure development. The network design problem is also useful in utilizing these resources efficiently in a more microscopic way, such as in a signal optimization problem. In addition, the problem can aid in transportation systems management (TSM).

When applied to highway investment decisions, the network design problem can aid in deciding on (1) the new links to be added to the network or (2) the expansion of existing links. The former called is a *discrete network design problem* and the latter a *continuous network design problem.*

For review of network design, see Magnanti and Wong (1984)[169] and Boyce (1984)[38]. Many different formulations of network design problems are possible based on specific objective functions and constraints. The following formulation, however, shows the basic formulation of the continuous network design problem.

$$P8.2) \quad \min_{y,x} \quad Z = \sum_a \int_0^{f_a} M_a(x)dx \tag{8.26}$$

$$\text{s.t.} \quad \sum_p X^{ijp} = T^{ij} \tag{8.27}$$

$$\sum_a G_a(y_a) \leq B \tag{8.28}$$

$$y, X^{ijp} \geq 0 \tag{8.29}$$

$$\text{where } M_a(x) = \frac{d(c_a(x) \cdot x)}{dx}$$

$$c_a(x) = t_0[1 + p^1(\frac{x}{s_a + y_a})^{p^2}]$$

$G_a(y)$: investment function

s_a : capacity of link a

y_a : capacity increase on link a

p^1, p^2 : parameters

B : total budget.

Capacity constraints for each link can also be added to the problem formulation. Notice that the formulation is essentially the same as the system-optimal equilibrium problem except for the budget constraint (8.28) and decision variable y. There are no constraints or features in the above formulation that require the resulting flows to be user-optimal flows. The user-optimal flow requirements have been left out mainly because of computational complexity. With the present form, **P8.1)** can identify the optimum investment levels that can minimize the total system travel cost within the budget constraint. As noted earlier, this is a supplier's view.

Despite its relative computational simplicity compared to the model that has user-equilibrium flow constraints, these kinds of system-optimal formulations have serious flaws, as pointed out by Braess (1968)[44]. It is possible that additional investments identified by this system optimum network design problem may increase total system cost when the flow is not required to be user-optimal. Thus arises Braess' paradox, which occurs because of the discrepancy between the system optimal flow pattern that is assumed in **P8.1)** and the user-optimal flow pattern, which is more likely to occur. One of the main thrusts for adding a user-optimal flow constraint in a network design problem is to avoid this paradox. The next section accounts for the procedures and details that are related to adding user-equilibrium flow constraints and, subsequently, the demand side in a network design model.

8.4 The Equilibrium Network Design Problem

An equilibrium network design problem refers to a network design problem where the flows are required to be user-equilibrium flows based on Wardrop's conditions (Wardrop 1952)[252]. The continuous equilibrium network design problem[5] (Abdulaal and LeBlanc 1979)[2] has the following form:

$$P8.3) \qquad \min_y Z = \sum_a [c_a(f_a(y), y_a) \cdot f_a(y) + \theta G_a(y_a)] \qquad (8.30)$$

$$\text{s.t.} \qquad y_a \geq 0 \qquad \forall a \qquad (8.31)$$

$$f_a(y) \quad \text{is a user equilibrium flow pattern.} \qquad (8.32)$$

The objective function (8.30) is basically the combination of (8.26) and (8.28).

[5]See LeBlanc (1975)[158] for a discrete version.

Following the seminal work of Abdulaal and LeBlanc (1979)[2], many researchers have tried to solve the continuous equilibrium network design problem exactly or heuristically. All suffer, however, from unknown functional forms of $f_a(y)$. To overcome this problem. Tan et al. (1979)[243] replaced constraint (8.32) with an equivalent set of path flow constraints and employed the Augmented Lagrangian method to solve the resulting problem. Abdulaal and LeBlanc (1979)[2] utilized the Hooke-Jeeves direct search method instead of working with the functional form of $f_a(y)$. Another approach, the iterative optimization-assignment algorithm proposed by Allsop (1974)[6] and Steenbrink (1974)[233], iterates between the assignment and optimization problem. This approach was later proven to be the exact algorithm for Cournot-Nash equilibrium (Tan et al., 1979[243]; Marcotte 1983[173]; Friesz and Harker 1985[102]). Recently, Suwansirikul et al. (1987)[241] devised a heuristic algorithm that uses a Bolzano-type search scheme to overcome the unknown functional form.

8.5 Bilevel Programming Formulations of ENDP

By noting that constraint (8.32) actually corresponds to the solution of the **P8.1**), the next two nonlinear bilevel programming formulations of the equilibrium network design problem can be obtained.

8.5.1 Model 1

The first model is formulated as follows:

$$
\begin{array}{lll}
P8.4)\quad \mathbf{U8.4)} & \min_y Z = \sum_a \int_0^{f_a(y)} M_a(x,y)dx + \theta \sum_a G_a(y_a) & (8.33)\\
\text{s.t.} & y_a \geq 0 & (8.34)\\
\text{where} & x \text{ solves} & \\
\mathbf{L8.4)} & \min_x \int_0^{f_a(y)} c_a(x,y)dx & (8.35)\\
\text{s.t} & \sum_p X^{ijp} = T^{ij} \qquad \forall i,j & (8.36)\\
& X^{ijp} \geq 0 \qquad \forall p, i, \text{ and } \; j & (8.37)\\
& f_a = \sum_i \sum_j \sum_p X^{ijp} \delta_a^{ijp} \qquad \forall a & (8.38)
\end{array}
$$

where

$$
\begin{array}{rcll}
M_a(f_a(y), y_a) & = & d[c_a(f_a(y), y_a) \cdot f_a(y)]/df_a(y) & (8.39)\\
c_a(f_a(y), y_a) & = & t_a[1 + p^1 \cdot [f_a(y)/(s_a + y_a)]^{p^2}]. & (8.40)
\end{array}
$$

$$
\begin{aligned}
M_a(f_a(y), y_a) &= \text{marginal cost on link } a \text{ with flow } f_a(y) \\
c_a(f_a(y), y_a) &= \text{average cost of link } a \text{ with flow } f_a(y) \\
p^1, p^2 &= \text{parameters} \\
G_a(y_a) &= \text{investment function for capacity increases} \\
y_a &= \text{capacity increase level for link } a \\
f_a(y) &= \text{total flow on link } a \text{ with capacity increase } y \\
\theta &= \text{coefficient converting construction cost to travel cost} \\
s_a &= \text{existing capacity of link } a \\
X^{ijp} &= \text{flow from } i \text{ to } j \text{ on path } p \\
\delta_a^{ijp} &= \begin{cases} 1 & \text{if link } a \text{ is included in path } p \text{ from } i \text{ to } j \\ 0 & \text{otherwise.} \end{cases}
\end{aligned}
$$

In this model, the planners at the upper level are assumed to make the decisions regarding investments for the transportation facilities (y). Specifically, the planners invest in certain links to increase their capacity such that the sum of total societal travel cost ($\int_0^{f_a(y)} M_a(x, y)dx$) and link improvement cost ($\theta \sum_a G_a(y_a)$) is minimized. The lower-level problem is a user-equilibrium route choice problem with flow conservation and nonnegativity constraints. The users at the lower level are assumed to follow the user-equilibrium principle of Wardrop (1952)[252] under the given facilities (y). Therefore, the link flows (f_a) are user-equilibrium flows.

8.5.2 Model 2

The second model, though similar in concept to the first one, has an explicit budget constraint in the upper-level problem. This model is structured as follows:

$$
\begin{aligned}
P8.5) \quad : \quad \mathbf{U8.5)} \quad & \min_y Z = \sum_a \int_0^{f_a(y)} M_a(x, y)dx && (8.41) \\
\text{s.t} \quad & \sum_a G_a(y_a) \le B && (8.42) \\
& y_a \ge 0 && (8.43) \\
\text{where} \quad & x \quad \text{solves} \\
\mathbf{L8.5)} \quad & \text{which is the same as } \mathbf{L8.4)}
\end{aligned}
$$

where B is the total available budget for link improvements or additions.

In the second model, the link improvement choice of the upper-level decision maker is explicitly limited by the total available budget for the project. Two models are equivalent if θ of **P8.4**) is used as the Lagrangian multiplier of the budget constraint (8.42). The adoption of two different models is solely for computational reasons. This point is explained later in the book.

Based on the description of section 8.2, the equilibrium flow constraint (equation (8.32)) can also be replaced by the nonlinear complementarity problem (NCP) or variational inequality (VI) formulation. Therefore, the following two models are also equivalent to **P8.4**).

$$P8.6) \quad : \min_{y} \quad Z = \sum_a \int_0^{f_a(y)} M_a(x,y)dx + \theta \sum_a G_a(y_a) \tag{8.44}$$

$$\text{s.t} \quad x \cdot F(x,u) = 0 \tag{8.45}$$

$$F(x,u) = \begin{bmatrix} c(x) - u^{ij} \\ \sum X^{ijp} - T^{ij} \end{bmatrix} \geq 0 \tag{8.46}$$

$$x = (x,u) \geq 0 \tag{8.47}$$

$$P8.7) \quad : \min_{y} \quad Z = \sum_a \int_0^{f_a(y)} M_a(x,y)dx + \theta \sum_a G_a(y_a) \tag{8.48}$$

$$\text{s.t} \quad c(x^*)(x - x^*) > 0 \tag{8.49}$$

$$\text{for all} \quad x \in \Omega \tag{8.50}$$

$$\Omega = \{x | \Lambda x = T, x > 0\}. \tag{8.51}$$

A bilevel programming model that used variation inequality (VI) formulation as a lower-level problem has been utilized by Marcotte (1986)[172] and Marcotte and Marquis (1988)[175].

8.6 Solution Algorithm: Introduction

There are about a dozen different solution algorithms for bilevel programming problems. None of those proposed, however, are directly applicable to solving the proposed nonlinear bilevel network design problem (Kim 1988)[142]. For nonlinear bilevel programming problems, either the Karush-Kuhn-Tucker approach or descent-type algorithm (or barrier method) is possible. The Karush-Kuhn-Tucker approach usually involves the branch and bound technique to resolve the complementarity condition (e.g., see Bard and Falk 1982)[20]. The branch and bound approach is notorious for its exponential growth of computational burden with an increasing number of variables. Therefore, descent algorithm is chosen as a basic approach in the research.

The existing descent algorithm, however, is not directly applicable to the proposed problem, since Wardropian user-equilibrium problems do not have unique user-equilibrium path flow patterns, even though the associated link flow pattern is unique (Aashitiani and Magnanti 1981[1]; Sheffi, 1985[224]; Tobin and Friesz 1988[245]). As was mentioned earlier, the lower-level problem of the proposed problem, **L8.4**, is a Wardropian user-equilibrium problem. Therefore, extra work is required to utilize the descent-type algorithm to solve the proposed nonlinear bilevel programming model of the equilibrium network design problem.

The proposed algorithm is based on a descent-type algorithm that relies on derivative information about the lower-level problem with respect to the upper-level decision variables in determining the search direction of the upper-level problem (de Silva 1978[79]; Kolstad and Lasdon 1986[147]). The derivative information is obtained by various methods. Once the derivative information of the lower-level variables with respect to the upper-level decision variables is available, many algorithms, including the Quasi-Newton method, are available to solve the upper-level problem. Thus, the calculation of a derivative is the most crucial part of a descent-type algorithm. Fiacco's (1976)[92] work has become the basis of calculating derivatives of a nonlinear problem, while Tobin's (1986)[246] or Dafermos's (1988)[71] work can be utilized for a problem formulated in variational inequality.

The next section briefly reviews the basic techniques for solving nonlinear problems and shows how the techniques for nonlinear programming problems have been utilized for the solution of bilevel programming problems. Based on these reviews, a direction is established for development of the solution algorithm for the bilevel network design problem. Derivative calculation procedures for the nonlinear programming problem is reviewed, followed by the development of derivative calculation procedures for a user-equilibrium problem. Finally, a descent-type algorithm is described for the nonlinear bilevel programming model of the equilibrium network design problem.

8.7 Solution Algorithms for Nonlinear Bilevel Problems

This section first summarizes the developments of solution algorithms for nonlinear programming problems, shedding light on how these algorithms can be utilized in solving bilevel programming problems.

The solution of unconstrained nonlinear programming problems

$$\min_{x} f(x) \tag{8.52}$$

$$x \in E^n \tag{8.53}$$

has been studied extensively. The solution algorithms can be grouped into two broad categories: one which requires derivatives of $f(x)$, and one which does not. The first category encompasses methods such as steepest descent, Newton's method, conjugate direction, and variable metric methods. The second category is usually referred to as the search method and includes the direct search method of Hooke and Jeeves (1962)[122], the random search method, and other variations.

In the following, Hooke and Jeeves' direct search method and Quasi-Newton method are briefly summarized to facilitate the expositions of the next section. Many mathematics textbooks (e.g., Himmelblau 1972[121], Avriel 1976[10], and Bertsekas 1982[31]) have detailed explanations of these methods.

The direct search method proposed by Hooke and Jeeves (1962)[122] and implemented by Wood (1960)[263] consists of two phases, an *exploratory search* around the base point and a *pattern search* in a direction selected for minimization. Figure 8.2 shows the flow chart of the method. Abdulaal and LeBlanc (1979)[2] used this search method for solving the continuous equilibrium network design problem.

The variable metric method is a certain class of descent-type algorithm that estimates either the Hessian matrix of the function to be minimized or its inverse. The Quasi-Newton method is one of the variable metric methods in which the Hessian matrix satisfies a *secant relation* (Avriel 1976[10]). Quasi-Newton methods generate the sequence of

$$x_{k+1} = x_k + \alpha_k d_k \tag{8.54}$$

$$d_k = -D_k \nabla f(x_k) \tag{8.55}$$

where D_k is a positive definite matrix, which makes equation (8.54) approximate Newton's method. It is adjusted during the computation. The

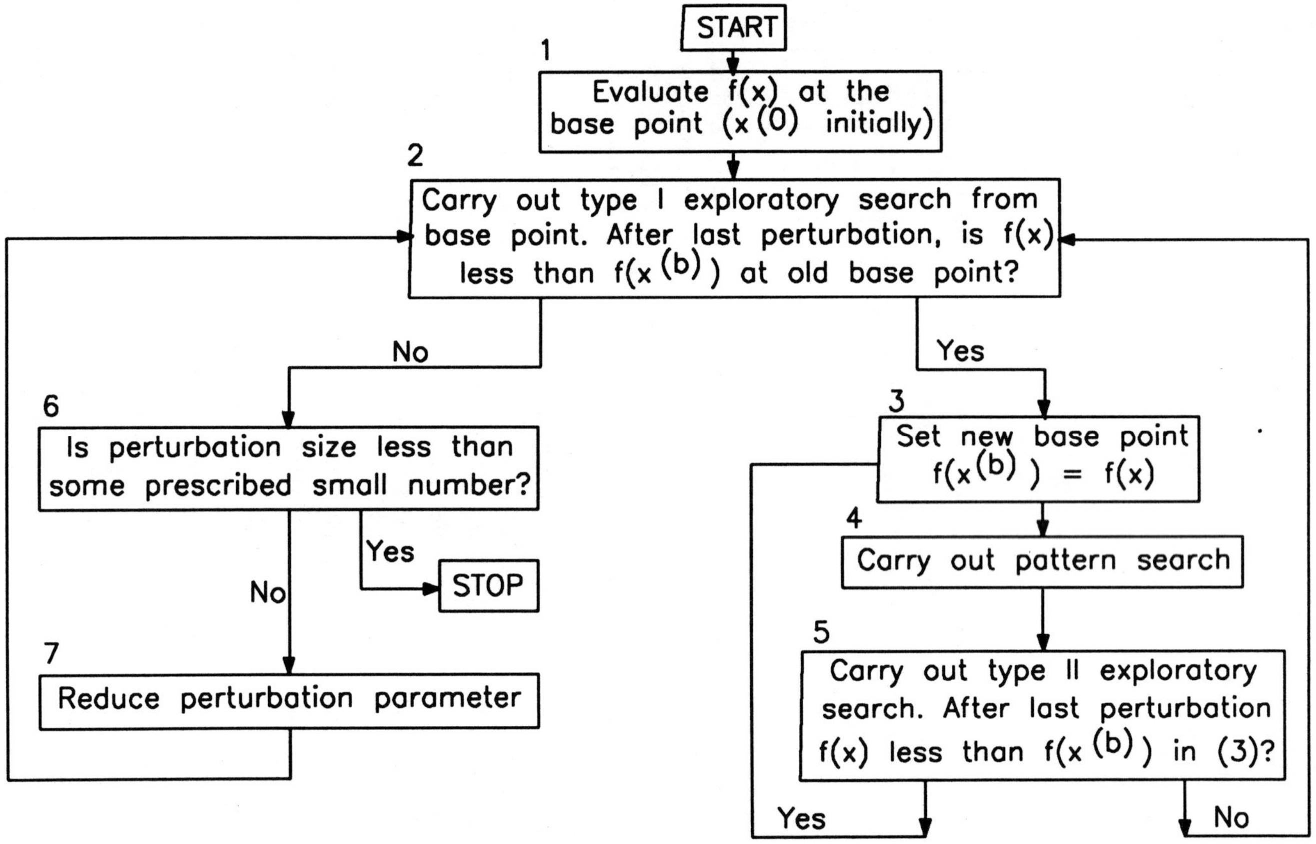

Source: Himmelblau 1972[121]

Figure 8.2: Flow Chart for Hooke-Jeeves Direct Search Method

stepsize α_k is determined by one of the stepsize rules, such as minimization, limited minimization, Armijo, Goldstein, or constant stepsize rules. The advantage of utilizing this Quasi-Newton method is that one can enjoy the fast convergence rate of Newton's method without computing the Hessian and its inverse. There are many variations of the methods, but the best general method is that of Broyden-Fletcher-Goldfarb-Shanno (BFGS) (Bertsekas 1982)[31]. This BFGS method is available in Harwell Library (AERE 1985)[3] as VA13 subroutine. The specific procedures of calculating descent directions are as follows:

$$\begin{aligned} D_{k+1} &= D_k + \frac{p_k \cdot p_k'}{p_k' \cdot q_k} - \frac{D_k \cdot q_k \cdot q_k' \cdot D_k}{q_k' \cdot D_k \cdot q_k} + \tau_k \nu_k \nu_k' \qquad (8.56)\\ \text{where} \quad & p_k = x_{k+1} - x_k \\ & q_k = \nabla f(x_{k+1}) - \nabla f(x_k) \\ & \nu_k = p_k - (1/\tau_k) \cdot D_k \cdot q_k \\ & \tau_k = q_k' \cdot D_k \cdot q_k / p_k' q_k. \end{aligned}$$

de Silva (1978)[79], and Kolstad and Lasdon (1986)[147] based their solution of the bilevel programming problem on this method, along with derivative information on the lower-level problems. Their upper-level problems were unconstrained problems. To solve the bilevel problem of finding which upper-level problem is constrained (as with this research), it is necessary to base the solution procedures on the algorithm for constrained nonlinear programming problems.

There are three broad categories of algorithms to solve the following constrained nonlinear programming problems (Bertsekas 1982)[31].

$$\min_x f(x) \qquad (8.57)$$

$$\text{s.t.} \quad h(x) = 0 \qquad (8.58)$$

$$g(x) \leq 0 \qquad (8.59)$$

$$f : R^n \longrightarrow R, h : R^n \longrightarrow R^m, g : R^n \longrightarrow R^r.$$

The first popular approach is descent-type. The method is based on the generation of iterative descent within the feasible set. Given a feasible point x_k, a *descent direction* d_k is chosen to satisfy the following two conditions:

1. Descent condition $\nabla f(x_k)' d_k < 0$.

2. For positive and sufficiently small α, $x_k + \alpha d_k$ is feasible.

A stepsize parameter α is determined to make the new feasible point $x_{k+1} = x_k + \alpha_x d_k$ satisfy $f(x_{k+1}) < f(x_k)$.

Generally, these kinds of methods are called *feasible direction methods*, or *projection methods*. There are many variations of these methods, depending on how the descent direction is decided. Most of the unconstrained algorithms using derivatives are special cases of the feasible direction method. The algorithm utilized in this research also belongs to this group.

The second approach is the penalty function approach. This method eliminates constraints through the use of penalty functions (see, for example, Fiacco and McCormick 1968)[93]. de Silva (1978)[79] utilized this penalty function approach to solve for the derivative information of the lower-level problem.

The third approach is the Lagrangian method. To solve the optimization problem, it solves the system of equations that constitute necessary conditions for optimality.

To summarize the above review, the descent-type method has been utilized mainly in solving bilevel problems. Futher, once the derivative of the lower level is obtained with respect to the decision variables of the upper level, many algorithms for the standard nonlinear programming problem can be utilized to solve the bilevel programming problem. Therefore, the calculation of derivatives of the lower-level problem with respect to the decision variables of the upper-level problem is the crucial part of successful application of the descent method for bilevel programming problems. The next section describes the method used to calculate the df/dy of **P8.4**).

8.8 Derivative Calculation

The following results are from Fiacco (1976)[92] and are presented here without proof for the sake of convenience.

$$\mathbf{P}(\epsilon) \quad \min_x \quad f(x,\epsilon) \tag{8.60}$$

$$\text{s.t.} \quad g_i(x,\epsilon) \geq 0 \tag{8.61}$$

$$h_j(x,\epsilon) = 0 \tag{8.62}$$

$$\text{where} \quad x \in E^n \text{ and } \epsilon \in E^k$$

Lemma 1 (Fiacco 1976.) *If the functions defining problem P(0) are twice continuously differentiable in a neighborhood of x^*, then x^* is a local isolated* *(locally unique) minimizing point of problem P(0) if there exists (Lagrange multiplier) vector $u^* \in E^m$ and $w^* \in E^p$ such that the first order Kuhn-Tucker conditions hold, i.e.,*

$$g_i(x^*, 0) \geq 0 \quad (i = 1, \ldots, m) \tag{8.63}$$
$$h_j(x^*, 0) = 0 \quad (j = 1, \ldots, p) \tag{8.64}$$
$$u_i^* g_i(x^*, 0) = 0 \quad (i = 1, \ldots, m) \tag{8.65}$$
$$u_i^* \geq 0 \quad (i = 1, \ldots, m) \tag{8.66}$$
$$\nabla L(x^*, u^*, w^*, 0) = \nabla f(x^*, 0) - \sum_{i=1}^{m} u_i^* \nabla g_i(x^*, 0) + \sum_{j=1}^{p} w_j^* \nabla h_j(x^*, 0) = 0 \tag{8.67}$$

and further if $y^T \nabla^2 L(x^*, u^*, w^*, 0) y > 0$ for all $y \neq 0$ such that

$$y^T \nabla g_i(x^*, 0) \geq 0 \quad \text{for all } i, \text{ where } g_i(x^*, 0) = 0, \tag{8.68}$$
$$y^T \nabla g_i(x^*, 0) = 0 \quad \text{for all } i, \text{ where } u_i^* > 0, \tag{8.69}$$
$$y^T \nabla h_i(x^*, 0) = 0 \quad (j = 1, \ldots, p). \tag{8.70}$$

Theorem 2 (Fiacco 1976.) *If*

1. *the functions defining $\mathbf{P}(\epsilon)$ are twice continuously differentiable in (x, ϵ) in a neighborhood of $(x^*, 0)$,*

2. *the second order sufficient conditions for a local minimum of $\mathbf{P(0)}$ hold at x^*, with associated Lagrange multipliers u^* and w^*,*

3. *the gradients $\nabla g_i(x^*, 0)$ (for i such that $g_i(x^*, 0) = 0$) and $\nabla h_j(x^*, 0)$ (all j) are linearly independent, and*

4. *$u_i^* > 0$ when $g_i(x^*, 0) = 0 \quad (i = 1, \ldots, m)$ (i.e., strict complementary slackness),*

then

1. *x^* is a local isolated minimizing point of $\mathbf{P(0)}$ and the associated Lagrange multipliers u^* and w^* are unique;*

2. *for ϵ in a neighborhood of 0, there exists a unique once continuously differentiable vector function $[x(\epsilon), u(\epsilon), w(\epsilon)]$ satisfying the second order sufficient conditions for a local minimum of problem $\mathbf{P}(\epsilon)$ such that $[x(0), u(0), w(0)] = (x^*, u^*, w^*)$ and, hence, $x(\epsilon)$ is a locally unique local minimum of $\mathbf{P}(\epsilon)$ with the associated unique Lagrange multipliers $u(\epsilon)$ and $w(\epsilon)$; and*

3. *strict complementarity (with respect to $u(\epsilon)$ and the inequality constraints) and linear independence of the binding constraint gradients hold at $x(\epsilon)$ for ϵ near 0.*

Assumption (2) implies the satisfaction of the Kuhn-Tucker first-order conditions at $(x, u, w, \epsilon) = (x^*, u^*, w^*, 0)$

$$\nabla L(x, u, w, \epsilon) = 0, \tag{8.71}$$

$$u_i g_i(x, \epsilon) = 0 \quad (i = 1, \ldots, m), \tag{8.72}$$

$$h_j(x, \epsilon) = 0 \quad (j = 1, \ldots, p) \tag{8.73}$$

With $(x, u, w) = [x(\epsilon), u(\epsilon), w(\epsilon)]$, equations (8.71) to (8.73) are satisfied for ϵ near 0 and can be differentiated with respect to ϵ to yield explicit expressions for the first partial derivatives of this vector function. It follows from this fact that the total derivative of equations (8.71) to (8.73) with respect to ϵ is zero for ϵ near 0 that

$$M(\epsilon) dt(\epsilon)/d\epsilon = N(\epsilon) \tag{8.74}$$

where $t(\epsilon) \equiv [x(\epsilon), u(\epsilon), w(\epsilon)]$, $M(\epsilon)$ is the Jacobian matrix of equations (8.71) to (8.73) with respect to (x, u, w), evaluated at $[t(\epsilon), \epsilon]$, and $N(\epsilon)$ is the negative of the Jacobian matrix of equations (8.71) to (8.73) with respect to ϵ, evaluated at $[t(\epsilon), \epsilon]$.

Since M is nonsingular for ϵ near 0,

$$dt(\epsilon)/d\epsilon = M^{-1}(\epsilon) N(\epsilon). \tag{8.75}$$

Corollary 1 (Fiacco 1986.) *Under the assumption of Theorem 2, a first-order approximation of $[x(\epsilon), u(\epsilon), w(\epsilon)]$ in a neighborhood of $\epsilon = 0$ is given by*

$$\begin{bmatrix} x(\epsilon) \\ u(\epsilon) \\ w(\epsilon) \end{bmatrix} = \begin{bmatrix} x^* \\ u^* \\ w^* \end{bmatrix} + (M^*)^{-1} N^* \epsilon + o(||\epsilon||) \tag{8.76}$$

where $(x^*, u^*, w^*) = [x(0), u(0), w(0)], M^* = M(0), N^* = N(0)$, *and* $M(\epsilon)$ *and* $N(\epsilon)$ *are defined as in (8.74).*

These theorems are the core of the sensitivity calculation procedures for nonlinear programming problems.

Direct application of Fiacco's (1976)[92] or Tobin's (1986)[246] work for obtaining the first-order derivative information (df/dy) of the lower-level problem **L7**, however, is not feasible. This follows because the Hessian matrix of the Lagrangian for **L7** in terms of link flow is positive definite if the congestion function is assumed to have a strictly increasing function of link flows, as was shown on page 127. This positive definiteness, however, does not hold in terms of path flows, as mentioned before. Thus, the solutions of **L7** in terms of path flow are not unique (see Sheffi 1985[224] for a simple example, and Tobin and Friesz 1988[245]). Unfortunately, there is no other way to reformulate **L7** without using path flow volume. This non-uniqueness of the solution prevents direct application of Fiacco's (1976)[92] or Tobin's (1986)[246] work, which require uniqueness of the solutions. The previous work of Dafermos and Nagurny (1984)[70] on sensitivity calculation cannot be used here because of their restrictive assumptions on the network structure.

The proposed alternative approach in this paper is a specialization of Tobin and Friesz's work (1988)[245]. Specifically, based on Tobin's earlier work (1986)[246] on variational inequality sensitivity analysis, Tobin and Friesz (1988)[245] considered a restricted problem and developed a procedure to calculate the derivative of link flows with respect to perturbation parameters for the user-equilibrium problem formulated in variational inequality. The proposed approach is a specialization of Tobin and Friesz's work(1988)[245] in the sense that its procedure is for nonlinear programming formulation ($L7$) and is confined to a specific algorithm, a convex combination algorithm otherwise known as the Frank-Wolfe algorithm, to solve **L7**. It is also shown in this research that the *restricted problem* of Tobin and Friesz can be developed by applying Fiacco's (1976)[92] and Kolstad and Lasdon's(1986)[147] work directly to the lower-level problem.

The Lagrangian of $L7$ in vector notation is

$$\mathcal{L}(\mathbf{x}, \pi, \mathbf{u}, \mathbf{y}) = \mathbf{c}(\mathbf{x}, \mathbf{y}) - \pi\mathbf{x} + \mathbf{u}(\mathrm{T} - \mathbf{x}).$$

Thus, the necessary conditions for solving L7 are that there exist $\mathbf{x}^*$, π^*, $\mathbf{u}^*$, $\mathbf{0}$ (for convenience, let $\mathbf{y} = \mathbf{0}$), which satisfy the following equations:

$$\mathbf{c}(\mathbf{x}^*, \mathbf{0}) - \pi^* - \Lambda^T \mathbf{u}^* \;=\; \mathbf{0} \qquad (8.77)$$

$$\pi^* \mathbf{x} = \mathbf{0} \quad (8.78)$$

$$\Lambda \mathbf{x}^* - \mathbf{T} = \mathbf{0} \quad (8.79)$$

$$\pi^* \geq \mathbf{0} \quad (8.80)$$

where Λ : an od/path incidence matrix.

This system of equations, however, does not meet the second order sufficient condition for a local isolated minimizing point (Lemma 1, page 138) because of the non-uniqueness of the solution, as was discussed on the page 127 (see also Tobin and Friesz 1988)[245].

If the convex combination method (Frank-Wolfe method) suggested by LeBlanc et al. (1975)[160] is utilized to solve the lower-level problem **L7**, the unique minimum cost paths between each origin/destination pair for each iteration of the solution process can be stored. This unique set of minimum cost paths gives the user-equilibrium flows. A unique path flow, which gives the unique link flows, can be identified by considering only this set of unique minimum cost paths. This set of unique path flows comprises an extreme point in the feasible region, which was discussed in Tobin and Friesz (1988)[245]. Thus, there is no need for a separate step to solve a linear programming problem in order to identify an extreme point in the feasible region, as Tobin and Friesz did.

Furthermore, only positive path flows are considered, because the convex combination algorithm assigns all origin/destination flows between O/D pairs on the minimum cost path at each iteration. (This is called an all-or-nothing assignment.) Therefore, nonnegativity constraint (8.37) will be nonbinding at the optimal solution and remain so for the perturbation of $\mathbf{y}$ in a neighborhood of $\mathbf{0}$. Fiacco (1983)[91] showed that nonbinding constraints can be accidentally deleted from the system of equations while the sensitivity information is obtained. Kolstad and Lasdon (1986)[147] went one step further and showed that the variables at their bounds can also be omitted from the sensitivity calculation. Using these results, the π^*, which is the Lagrange multiplier of the nonbinding constraint (8.37), can be deleted without affecting the outcome of the sensitivity calculation. Thus, the system of equations reduces to

$$\mathbf{c}(\mathbf{x}^*, \mathbf{0}) - \Lambda^{\mathrm{T}} \mathbf{u}^* = \mathbf{0} \quad (8.81)$$

$$\Lambda \mathbf{x}^* - \mathbf{T} = \mathbf{0}. \quad (8.82)$$

This is the same as the *restricted problem* of Tobin and Friesz (1988)[245]. They showed that this system of equations meets sufficient conditions for a locally unique solution (Theorem 3 of Tobin 1986)[246]. In fact, the optimality conditions for a variational inequality problem are identical with those for the nonlinear program (Theorem 2, page 139). Thus, it can be trivially shown that the above system of equations also satisfies the four constraint qualifications of Theorem 2 (page 139). Once a system of equations is obtained that meets all the conditions for sensitivity analysis, Fiacco's approach (1976)[92] for these equations to calculate df/dy can be readily applied.

With $\mathbf{z} = (\mathbf{x}, \mathbf{u}) = [\mathbf{x}(\mathbf{y}), \mathbf{u}(\mathbf{y})]$, equations (8.81) and (8.82) are also satisfied for $\mathbf{y}$ near $\mathbf{0}$ and can be differentiated with respect to $\mathbf{y}$ to yield explicit $\mathbf{dx/dy}$ and $\mathbf{du/dy}$. Because the total derivative of equations (8.81) and (8.82) with respect to $\mathbf{y}$ is $\mathbf{0}$ for $\mathbf{y}$ near $\mathbf{0}$, the $\mathbf{dz} = (\mathbf{dx/dy}, \mathbf{du/dy})$ is the solution of the following system of linear equations (8.74):

$$\mathbf{M}(\mathbf{y}) \cdot \mathbf{dz/dy} = \mathbf{N}(\mathbf{y}) \tag{8.83}$$

where

M the Jacobian matrix of equations (8.81) and (8.82) with respect to (x, u) evaluated at $[z(y), y]$ and

N the negative of the Jacobian matrix of equations (8.81) and (8.82) with respect to y evaluated at $[z(y), y]$.

More precisely, the following system of linear equations is to be solved to get $dz(y)/dy$:

$$\begin{bmatrix} \nabla_x c(\mathbf{x}^*, \mathbf{0}) & -\Lambda^T \\ \Lambda & \mathbf{0} \end{bmatrix} \cdot \begin{bmatrix} dx(y)/dy \\ du(y)/dy \end{bmatrix} = - \begin{bmatrix} \nabla_y c(\mathbf{x}^*, \mathbf{0}) \\ \mathbf{0} \end{bmatrix} \tag{8.84}$$

where

$$\begin{aligned} \nabla_x c(x^*, 0) &= \Delta^T \nabla_f c(f^*, 0) \Delta = \partial(\text{path cost})/\partial(\text{path volume}) \\ \nabla_y c(x^*, 0) &= \Delta^T \nabla_y c(f^*, 0) = \partial(\text{path cost})/\partial(\text{capacity increase}) \\ \Delta &: \text{link/path incidence matrix.} \end{aligned} \tag{8.85}$$

The first order derivative of link flow with respect to $\mathbf{y}$ at $\mathbf{y} = \mathbf{0}$ is

$$[df(y)/dy] = \Delta \cdot [dx(y)/dy]. \tag{8.86}$$

Once $df(y)/dy$ is known, dZ/dy for **P7** can be calculated as follows:

$$\begin{aligned} dZ/dy &= \sum_a c_a(f_a(y), y) \cdot \partial f_a(y)/\partial y + \sum_a (\partial c_a(f_a(y), y)/\partial y \\ &\quad + \partial c_a(f_a(y), y)/\partial f_a \cdot \partial f_a/\partial y) \cdot f_a(y) + dG(y)/dy. \end{aligned} \tag{8.87}$$

8.9 Developing Solution Algorithms: A New Approach

Many different methods of solving the upper-level problem can be utilized to solve the overall problem once we know the derivative of the lower-level problem with respect to the decision variables of the upper-level problem. The basic idea is to solve the bilevel programming problem as if it were a single-level nonlinear programming problem. After the solution process starts, it successively generates points based on the gradients of the upper-level objective functions with respect to the upper-level decision variables. The derivative of the lower-level problem is utilized in the calculation of the gradients of the upper-level objective function. If the upper-level problem of the bilevel programming problem has constraints, the nonlinear programming algorithms for a constrained nonlinear programming problem can be utilized. If no constraints are involved with the upper-level problem, nonlinear programming algorithms for the unconstrained problem can be adopted. In summary, the generic descent-type algorithm for the overall problem can be stated in the following steps:

Step 0. Initialize capacity increase y.

Step 1. Solve **L7** for given $\mathbf{y}$ with the Frank-Wolfe algorithm, obtain link/path incidence matrix (Δ), O/D path incidence matrix (Λ), and optimal link flows given $\mathbf{y}$ ($\mathbf{f}^*$).

Step 2. Solve the system of equations (8.84) and (8.86) to obtain dx/dy and $df/dy \quad (= \Delta \cdot dx/dy)$

Step 3. Calculate dZ/dy, equation (8.87), and obtain the next $\mathbf{y}$ value for evaluation according to a specific search scheme. If convergence is obtained $\longrightarrow$ STOP, otherwise go to step 1.

The upper-level problem of P7) is considered a minimization problem subject to simple constraints that are bound to the variables. The problem P8) can be considered either a linearly- or nonlinearly-constrained minimization problem, depending on the functional form of the investment function ($G_a(y_a)$).

Two different approaches are taken to solve **U7**. The first approach for P7) is a modification of the Broyden-Fletcher-Goldfarb-Shanno (BFGS) Quasi-Newton method (see, for example, Bertsekas 1982)[31] for an unconstrained problem to accommodate the nonnegativity constraints. When the constraints are violated during the search process, a new perturbed point is generated, making the violated constraints active. The second approach for P8) utilizes one of many constrained minimization methods, that use the feasible direction method, as was discussed in section 4.2.

The VA13 subroutine, which is the BFGS routine of Harwell Library, is utilized with the above modification for the first approach; and VE03, which is based on Fletcher (1971)[96], is adopted for the second approach. The problem P8) can be solved by either VE03 or VF03, depending on the functional form of the investment function ($G_a(y_a)$). The VF03 employs the watchdog technique (Powell 1982[204]; Chamberlain et al. 1982[60]). The algorithm can be regarded as an extension of a variable metric method for unconstrained optimization to the constrained case (Harwell 1985)[3]. The algorithm utilized in the VF03 is described in more detail in Powell (1982)[204]. It also utilizes directional derivatives and the Armijo stepsize selection rule (Armijo 1966)[7]. No formal proof of convergence has been given here, since the BFGS method (VA13 routine) is proven to guarantee convergence to a local solution under mild assumptions (Bertsekas 1982)[31]. The VA13 routine ensures the positive definiteness of the approximated inverse of the Hessian matrix at each iteration. Furthermore, the watchdog technique actually forces convergence. de Silva (1978)[79] presented a proof of the local convergence of this descent-type approach when the penalty function method is used to solve the lower-level problem of the convex problem. equation (8.84) is solved either by the LU factorization method([82],[266],and [84]) or the iterative method[218]. For a large-scale problem, a sparse matrix storage scheme should be used.

The performance of the algorithm must be tested before it can be applied to a larger problem. To test the performance of the algorithm, it is applied to two small-scale equilibrium network design problems with known solutions. The next section reports the computational experience for the example problems with the algorithm proposed in this research.

8.10 Testing the Solution Algorithm

Two example networks are taken from Suwansirikul et al. (1987)[241]. These examples are useful, because computation results obtained from other algorithms are provided for comparison purposes. Both P7 and P8 are solved with the sample data.

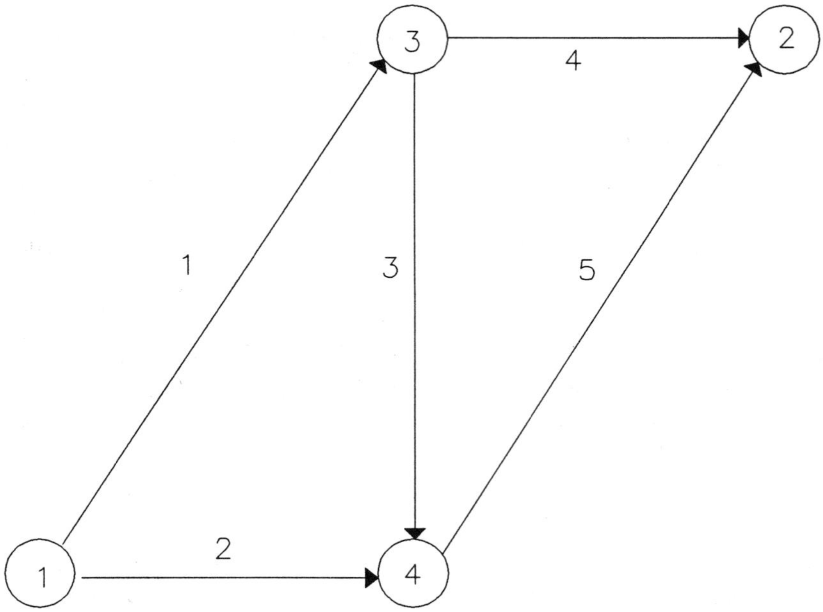

Figure 8.3: Test Network 1

8.10.1 Example 1

The first sample network is shown in Figure 8.3, with associated data presented in Table 8.1.

This problem has one O/D pair, four nodes and five links. Table 8.2 shows the solutions of P7 with various traffic demands, along with the solutions obtained with other algorithms from past studies.

The same convergence criteria as reported in Suwansirikul et al. (1987)[241] are used; but the computational resource requirement of the proposed algorithms in terms of CPU time cannot be compared, because none is available with which to compare it. Table 8.3 compares the computational burden of the proposed algorithm against other algorithms in terms of the number of Frank-Wolfe iterations required for obtaining solutions.

The computational results of the proposed algorithm for P8.5 formulation with different total budget levels are shown in Table 8.4.

From the above discussions, one can easily conclude that the proposed algorithm works well in terms of accuracy and computational burden. The algorithm gives the same solutions as those obtained by Tan (Tan et al. (1979)[241], which is regarded as an exact method of solving an equilibrium

Link	A_a	B_a	K_a	d_a
1	4.0	0.60	40.0	2.0
2	6.0	0.90	40.0	2.0
3	2.0	0.30	60.0	1.0
4	5.0	0.75	40.0	2.0
5	3.0	0.45	40.0	2.0

$$\begin{aligned} c_a(f_a, y_a) &= A_a + B_a(f_a/(K_a + y_c))^4 \\ Z(y) &= \sum_a (c_a(f_a, y_a) \cdot f_a + 1.5 \cdot d_a \cdot y_a^2) \end{aligned}$$

Table 8.1: Data for Test Network 1

network design problem. Furthermore, the algorithm consistently requires fewer Frank-Wolfe iterations to reach the optimum solutions.

8.10.2 Example 2

The second example network and data are shown in Figure 8.4 and Table 8.5. This example has six nodes (two are O/D zones) and sixteen links. Tables 8.6 and 8.7 show the computation results for the problem. The existence of many local solutions are reconfirmed. The computer codes used in this research find the local solution that is closest to the starting solution. Notice that each solution algorithm gives not only different objective function values, but also a different set of link improvements. The solution algorithm proposed in this research again produced the optimum solution consistently and with fewer Frank-Wolfe iterations.

8.11 Concluding Remarks

In this chapter, two nonlinear bilevel programming models of the continuous equilibrium network design problem have been formulated based on the descriptions about the network equilibrium, network design problem, and continuous equilibrium network design problem. More specifically, the network equilibrium problem has been revisited. Based on the uniqueness of the solution to the network equilibrium problem, desirable characteristics

Demand	Link	MINOS	H-J	EDO	BDA
100	y_1	1.34	1.25	1.31	1.34
	y_2	1.21	1.20	1.19	1.21
	y_3	0.00	0.00	0.06	0.00
	y_4	0.97	0.95	0.94	0.97
	y_5	1.10	1.10	1.06	1.08
	Z	1200.58	1200.61	1200.64	1200.58
150	y_1	6.05	5.95	5.98	6.04
	y_2	5.47	5.65	5.52	5.46
	y_3	0.00	0.00	0.02	0.00
	y_4	4.64	4.60	4.61	4.64
	y_5	5.27	5.20	5.27	5.27
	Z	3156.21	3156.38	3156.24	3156.21
200	y_1	12.78	13.00	12.86	12.98
	y_2	11.73	11.75	12.02	11.73
	y_3	0.00	0.00	0.02	0.00
	y_4	10.34	10.25	10.33	10.34
	y_5	11.74	11.75	11.77	11.74
	Z	7086.12	7086.21	7086.45	7086.13
300	y_1	28.45	28.44	28.11	28.45
	y_2	25.73	25.75	26.03	25.73
	y_3	0.00	0.00	0.01	0.00
	y_4	23.40	23.44	23.39	23.40
	y_5	26.57	26.56	26.58	26.57
	Z	21209.90	21209.91	21210.54	21209.90

H-J Hooke-Jeeves algorithm of Abdulaal and LeBlanc 1979[2]

EDO Equilibrium Decomposed Algorithm of Suwansirikul et al. 1987[241]

BDA Bilevel Descent Algorithm proposed in this paper

MINOS Tan et al. 1979[243]

H-J, EDO, and MINOS results are from Suwansirikul et al. 1987[241].

Table 8.2: Comparison of Results for Test Network 1

Demand	Algorithm	No. of Equilibrium Problem Solved	No. of Frank-Wolfe Iteration
100	H-J	11	27
	EDO	8	24
	BDA	5	17
150	H-J	14	37
	EDO	11	29
	BDA	6	19
200	H-J	20	56
	EDO	12	32
	BDA	7	19
300	H-J	28	79
	EDO	14	38
	BDA	10	25

H-J Hooke-Jeeves algorithm of Abdulaal and LeBlanc 1979[2]

EDO Equilibrium Decomposed Algorithm of Suwansirikul et al 1987[241]

BDA Proposed Bilevel Descent Algorithm

H-J and EDO results are from Suwansirikul et al. 1987[241].

Table 8.3: Comparison of Computational Burdens for Test Network 1

Budget	Link	Improvement	No. of Iteration	No. of F-W Iteration
100	y_1	3.28	6	12
	y_2	2.97		
	y_3	0.00		
	y_4	2.45		
	y_5	2.79		
	Z	1141.15		
50	y_1	2.34	7	15
	y_2	2.11		
	y_3	0.00		
	y_4	1.72		
	y_5	1.95		
	Z	1161.28		
15	y_1	1.29	8	20
	y_2	1.16		
	y_3	0.00		
	y_4	0.93		
	y_5	1.06		
	Z	1185.74		

Table 8.4: Computational Results for P8.5 Formulation

Link	A_a	B_a	K_a	d_a
1	1.0	10.0	3.0	2.0
2	2.0	5.0	10.0	3.0
3	3.0	3.0	9.0	5.0
4	4.0	20.0	4.0	4.0
5	5.0	50.0	3.0	9.0
6	2.0	20.0	2.0	1.0
7	1.0	10.0	1.0	4.0
8	1.0	1.0	10.0	3.0
9	2.0	8.0	45.0	2.0
10	3.0	3.0	3.0	5.0
11	9.0	2.0	2.0	6.0
12	4.0	10.0	6.0	8.0
13	4.0	25.0	44.0	5.0
14	2.0	33.0	20.0	3.0
15	5.0	5.0	1.0	6.0
16	6.0	1.0	4.5	1.0

$$\begin{aligned} c_a(f_a, y_a) &= A_a + B_a(f_a/(K_a + y_a))^4 \\ Z(y) &= \sum_a (c_a(f_a, y_a) \cdot f_a + d_a \cdot y_a) \end{aligned}$$

Table 8.5: Data for Test Network 2

Demand	Link	MINOS	H-J	EDO	BDA
7.5	y_3	0.0	0.0	0.0	0.0
	y_6	5.0	0.3	1.84	0.0
	y_7	0.0	0.3	0.02	0.0
	y_15	1.33	0.1	0.02	0.76
	y_16	0.0	0.3	1.84	0.0
	Z	92.1	90.10	92.41	91.92
15	y_3	0.0	1.2	0.13	0.0
	y_6	6.58	3.0	6.26	5.11
	y_7	0.0	0.0	0.0	0.0
	y_15	7.01	3.0	0.13	0.0
	y_16	0.22	2.80	6.26	5.71
	Z	211.25	215.08	201.84	202.18

H-J Hooke-Jeeves algorithm of Abdulaal and LeBlanc 1979[2]

EDO Equilibrium Decomposed Algorithm of Suwansirikul et al. 1987[241]

BDA Bilevel Descent Algorithm proposed in this paper

MINOS Tan et al. 1979[243]

H-J, EDO, and MINOS results are from Suwansirikul et al. 1987[241].

Table 8.6: Comparison of Results for Test Network 2

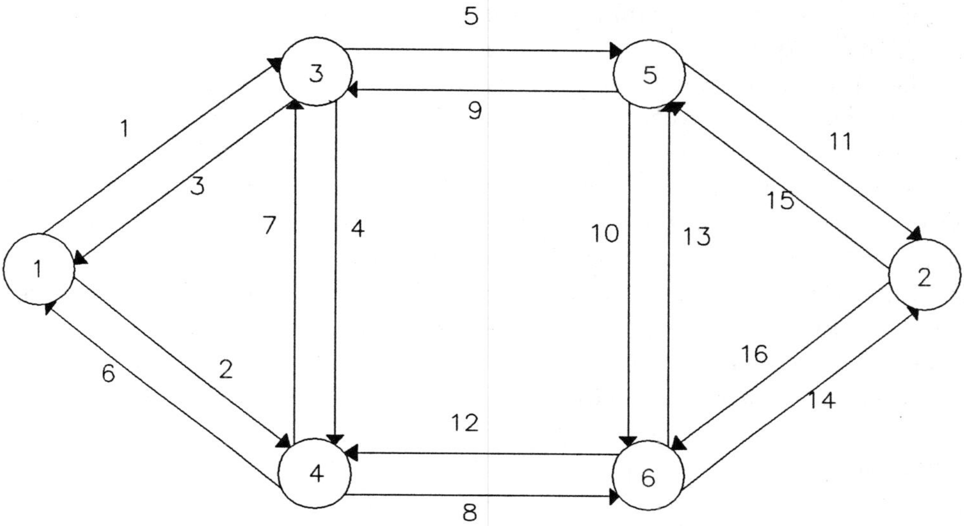

Figure 8.4: Test Network 2

Demand	Algorithm	No. of Equilibrium Problem Solved	No. of Frank-Wolfe Iteration
7.5	H-J	39	115
	EDO	11	57
	BDA	3	11
15	H-J	54	166
	EDO	10	71
	BDA	8	66

H-J Hooke-Jeeves algorithm of Abdulaal and LeBlanc 1979[2]

EDO Equilibrium Decomposed Algorithm of Suwansirikul et al 1987[241]

BDA Proposed Bilevel Descent Algorithm

H-J and EDO results are from Suwansirikul et al. 1987[241].

Table 8.7: Comparison of Computational Burdens for Test Network 2

of the link congestion function are set forth that will be used in the development of the link congestion function in the chapter that follows. Also, utilization of the equilibrium network design problem that defies the occurrence of the Braess' paradox has been supported. Finally, the application of the bilevel programming problem to formulating the equilibrium network design problem is advocated and two bilevel network design problems are formulated.

A solution algorithm is also developed to solve the nonlinear bilevel programming model of the equilibrium network design problem. The algorithm is developed using a descent-type approach that solves problems iteratively. It starts from an initial point and successively generates points that converge to a solution. The algorithm uses the gradients of the upper-level objective function with respect to the capacity improvements to generate next points. The derivative of the link flow with respect to the capacity improvements is utilized in the calculation of the gradients of the upper-level objective function. Also, a procedure to calculate the derivative of the link flow with respect to the capacity improvements is introduced and operationalized in this chapter.

The performance of the algorithm has been demonstrated with two small-sized equilibrium network design problems. For the small problems considered, the algorithm performs even better than heuristic algorithms when the number of Frank-Wolfe iterations is used as a proxy for the algorithmic efficiency.

Chapter 9

Model Application I: Measurement of Congestion Functions for Korean Highways

9.1 Introduction

Travel cost, or in a more general sense, travel impedance, on a given highway link usually increases as the traffic flow increases.[1] As traffic flow increases beyond a certain level, congestion takes place. In modeling a highway network with congestion effect, it is necessary to have a correct congestion function to account for increases in travel cost for the given highway environment. This highway environment includes driver characteristics, roadway conditions, and roadside activities. Different countries with distinctive demographic, economic, cultural, and behavioral characteristics might have unique congestion functions for their own environments.

The congestion function developed by the Bureau of Public Roads (BPR 1964)[52] of the United States has been used in Korea. To date there has not been much effort directed toward calibrating the parameters of the function. Korea, however, has traffic characteristics distinctive from those of the United States. Thus, it is desirable for Korea to have its own congestion

[1]This chapter, in its revised form, was written with Sunduck Suh and Chang Ho Park is forthcoming in *Transportation Research.*

function that can appropriately represent appropriately the Korean highway environment. Any attempt to model the Korean highway system without using a suitable congestion function might result in inappropriate solutions, because most modeling activities are crucially based on link costs, and it is the congestion function that furnishes those link costs.

This congestion function is calibrated for the Korean highway environment in this chapter. Toward this end, reviews on the congestion function are described in section 2. Because the definition of capacity plays a vital role in the development of congestion functions, a brief description of highway capacity is included in section 3. Also included in section 3 is a review of current highway capacity estimation in Korea. A bilevel programming model for estimating the parameters of the highway congestion function is introduced in section 4. In section 5, the bilevel programming model is applied to the Korean highway system to calibrate parameters of the congestion function that are appropriate for the Korean highway system. A summary and conclusions follow.

9.2 Link Congestion Function: An Overview

For any given link, it has been generally observed that speed decreases as the flow increases up to the capacity of the link. In the congested portion of the curve, both flow and speed decrease (Figure 9.1(a)) while travel time increases (Figure 9.1(b)). This function, which represents the relationship between flow and travel time on a link, has many different names: *capacity function, capacity restraint function, link congestion function, link performance function, congestion function,* and *link cost function.* In this book the function is referred to as a link congestion function or congestion function.

The link congestion function becomes important in the assignment procedure (also called the *route choice procedure*) because it is this function that determines the congestion costs of a link. Without an accurate link congestion function, it is impossible to accurately model the user's route choice behavior, which is based on a traveler's perception of travel time and cost.

The actual congestion function, such as that shown in Figure 9.1(b), has limitations, because this congestion function will not give a unique flow solution. A congestion function has to be a monotonically increasing function of flow for a mathematical programming assignment model to have a unique flow solution (see, for example, Sheffi 1985)[224]. Therefore, most research

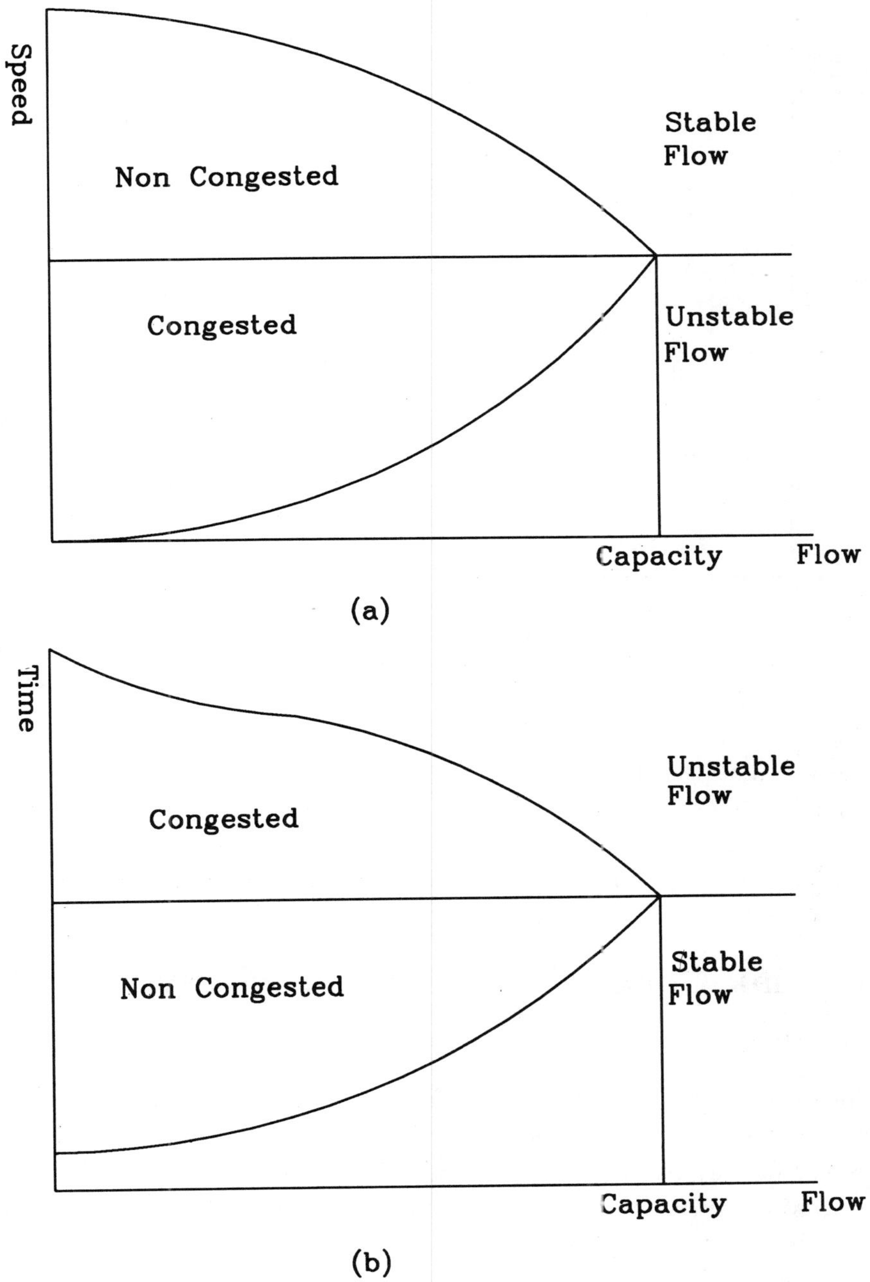

Figure 9.1: Speed Versus Flow (a) and Time Versus Flow (b) Relationship

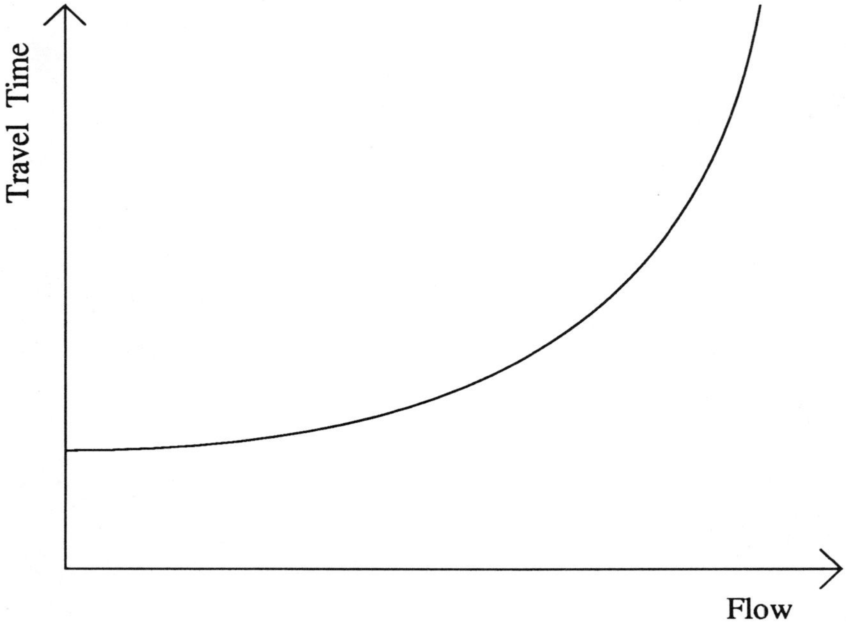

Figure 9.2: A Typical Convex Congestion Function

typically utilizes a convex link congestion function (Figure 9.2).

One notable exception is the work of Okutani (1984)[198], who utilized two convex functions to approximate the original shape (Figure 9.1(b)) in the assignment procedure. Multiple local solutions have resulted, however.

9.3 Link Congestion Function: Measurement Issues

Although many different congestion functions have been proposed and utilized, there has been no consensus about the type of congestion function necessary for any particular links.

Branston (1976)[46] classified two main approaches taken in defining the link congestion function, the *mathematical function approach* and the *theoretical approach.* In the mathematical function approach, a simple mathematical function replicating observed data is devised. Because of its simplicity, it is usually hard to incorporate network or link characteristics. In the theoretical approach, network characteristics such as signal spacing, signal settings, and/or link characteristics are well represented based on the queuing theory. Therefore, the resulting congestion function becomes more complex than the mathematical function approach.

The earlier research development of the congestion function is summarized, based largely on Branston's work (1976)[46] and is listed next for the sake of completeness. The detailed description of most of the work mentioned here can be found in Branston (1976)[46].

9.3.1 Mathematical Function Approach

1. Irwin, Dodd, and Von Cube (1961)[124] proposed a function of two straight line segments:

$$\begin{aligned} T &= T_a + \alpha(Q' - C'_p) \quad \text{for } Q' < C'_p \\ T &= T_a + \beta(Q' - C'_p) \quad \text{for } Q' \geq C'_p \qquad (9.1) \\ \text{where } T_a &= T_0 + \alpha C'_p \\ Q &= \text{flow on a link} \\ T &= \text{travel time at flow } Q \\ T_0 &= \text{travel time at zero flow} \\ C_p &= \text{practical capacity} \\ \alpha, \beta &= \text{parameters} \\ Q' &= \text{flow on a link per lane} \\ C'_p &= \text{practical capacity per lane.} \end{aligned}$$

This function was applied to the Toronto network. This two-line segment function later evolved into a three-line segment function later by Irwin and Von Cube (1962)[125]:

$$\begin{aligned} T &= T_a + \alpha(Q' - C'_p) \quad \text{for } Q' < C'_p \\ T &= T_a + \beta(Q' - C'_p) \quad \text{for } C'_p \leq Q' \leq C' \\ T &= T_b + \gamma(Q' - C') \quad \text{for } Q' > C'_s \qquad (9.2) \end{aligned}$$

where

$$\begin{aligned} T_a &= T_0 + \alpha C'_p \\ T_b &= T_a + \beta(C' - C'_p) \\ C' &: \text{Level of Service (LOS) E capacity.} \end{aligned}$$

It has been reported that the predicted flows from the assignment procedure with these congestion functions agreed reasonably well with the observed data. These functions, however, cannot be used in the mathematical programming assignment problem because of the discontinuities at C'_p and C'. The development of these two-line segment and three-line segment linear congestion functions parallels more recent interests on piece-wise linearization (approximation) of nonlinear congestion functions (equation 9.6) (Morlok et al. 1974[185]; LeBlanc and Boyce 1986[159]; Ben-Ayed 1988[27]).

2. A nonlinear function used in the Detroit Area Transportation Study was proposed by Smock (1962)[229]:

$$T = T_0 \exp[Q/C]. \tag{9.3}$$

3. Mosher's (1963)[187] logarithmic link function is defined as follows:

$$\begin{aligned} T &= T_0 + \ln(\alpha) - \ln(\alpha - Q) && (9.4) \\ \text{where} \quad & Q \leq \alpha. \end{aligned}$$

A hyperbolic capacity function is defined as

$$\begin{aligned} T &= \beta - \frac{\alpha(T_0 - \beta)}{Q - \alpha} && (9.5) \\ \text{where} \quad & Q \leq \alpha. \end{aligned}$$

4. In 1964, the Bureau of Public Roads (BPR 1964)[52] developed the most widely used congestion function, usually known as the BPR formula.

$$\begin{aligned} T &= T_0[1 + \alpha(Q/C_p)^\beta] && (9.6) \\ \text{where} \quad & \alpha = 0.15, \quad \beta = 4. \end{aligned}$$

5. In the Pittsburgh Area Transporation Study, Soltman (1965)[230] developed another nonlinear congestion function.

$$\begin{aligned} T &= T_0 2^{Q/C_p} && (9.7) \\ \text{where} \quad & Q/C_p \leq 2. \end{aligned}$$

6. The Traffic Research Corporation (1966)[247] used the following function for the Winnipeg area:

$$T = \alpha + \beta(Q' - \gamma) + \sqrt{[\beta^2(Q' - \gamma)^2 + \delta]}. \tag{9.8}$$

7. Equation (9.7) was generalized by Overgaard (1967)[199] as

$$T = T_0 \alpha^{(Q/C_p)^\beta}. \tag{9.9}$$

Equation (9.7) is the special case of equation (9.9) when $\alpha = 2$ and $\beta = 1$.

8. Dafermos (1968)[69] proposed the following function:

$$\begin{aligned} T_a &= \alpha_a\{\beta_a/(\beta_a + Z_a)\}^{u_a} X_a^2 + B_a X_a \qquad (9.10)\\ \text{where} \quad & \alpha_a, \beta_a > 0 \\ & 0 \le u_a \le 1 \\ Z_a : \quad & \text{capacity added to link } a \\ X_a : \quad & \text{flow on link } a \\ B_a(\cdot) : \quad & \text{function.} \end{aligned}$$

9. In 1974, Steenbrink (1974)[233] substituted the practical capacity of BPR formula with LOS E capacity (C) and calibrated α and β for Dutch environment.

$$\begin{aligned} T &= T_0[1 + \alpha(Q/C)^\beta] \qquad (9.11)\\ \alpha &= 2.62 \\ \beta &= 5. \end{aligned}$$

9.3.2 Theoretical Approach

1. Campbell, Keefer, and Adams (1959)[54] reported a function known as CATS (Chicago Area Transportation Study) based on a theoretical approach:

$$\begin{aligned} T &= T_0 \quad \text{for } \frac{Q}{C_s} \le 0.6 \\ T &= T_0 + \alpha\left(\frac{Q}{C_s} - 0.6\right) \quad \text{for } \frac{Q}{C_s} > 0.6. \qquad (9.12) \end{aligned}$$

2. Another popular link congestion function belonging to this category is that of Davidson (1966)[78]:

$$T = t(1 + J\frac{C}{1-C}) \qquad (9.13)$$

$$C = Q/S \qquad (9.14)$$

where

S : Saturation parameter

J : Delay parameter

t : zero flow time.

This function attracted interests from many researchers (Boyce, Janson, and Eash 1981[40]; Daganzo 1977[73]).

3. Wardrop(1968)[251] developed a network congestion function that shows the relationship between overall travel speed in the network and flow. The corresponding link capacity congestion function is

$$T = \frac{T_0}{(1-\gamma Q)} + \frac{\alpha\beta}{(\alpha - Q)D} \qquad (9.15)$$

where

D : link length

$$\alpha > C$$

$$\gamma < \frac{1}{C}.$$

9.3.3 Current Practice in Korea

The BPR formula has been the *de facto* standard congestion function for Korea. The function has been widely used even without any efforts to calibrate the parameters (α, β) for the Korean environment. Recently, some efforts to calibrate the parameters for different congestion functions have been reported. Choi (1987)[62] tried to calibrate the J factor of the Davidson formula (equation (9.13)) for the Korean urban environment. Lee (1987)[161] attempted to calibrate the BPR formula's (equation (9.6)) parameter for the Korean environment with ad hoc procedures. Both Choi and Lee tried several numbers for the parameters for the assignment procedure and chose the best among the numbers that yielded the best statistics. Even though both works suffer from lack of data[2] and statistical insignificance,[3] these are the only efforts thus far for studying the Korean environment. The work in this paper is systematic and presents a large-scale endeavor to calibrate parameters of a BPR-type (equation (9.6)) link congestion function for the Korean environment. As can be identified from the review of previous works on congestion functions, there are two different capacities involved in the mathematical programming approach for congestion function development. The first capacity is the *practical capacity.* The 1950 Highway Capacity Manual (HCM) refers to the practical capacity (BPR 1950)[51]. The second capacity, the Level of Service (LOS) E capacity, was introduced in the 1965 HCM (TRB 1965)[120] and reinforced in the 1985 HCM (TRB 1985)[248].

The next section briefly summarizes the changes in the definitions of capacities from the 1950 HCM to the 1985 HCM. A description of the current Korean practices on highway capacity calculations follows.

9.4 Highway Capacity

9.4.1 An Overview

The concept of highway capacity has changed twice since 1950. In that year, the 1950 HCM (BPR 1950)[51] defined three different levels of highway capacity concepts: *basic capacity, possible capacity*, and *practical capacity,* noting that the word *capacity* itself is just a generic expression for the ability of a roadway to accommodate traffic. Practical capacity is defined as

> *the maximum number of vehicles that can pass a given point on a roadway or in a designated lane during one hour without the traffic density being so great as to cause* unreasonable *delay, hazard, or restriction to the drivers' freedom to maneuver under the prevailing roadway and traffic conditions* (BPR 1950)[51].

The maximum practical capacity was 1,500 passenger cars per lane per hour for urban areas and 1,000 passenger cars per lane per hour for multi-lane highways in rural areas. Calculations for two-lane highways used 1,500 passenger cars per hour for urban areas and 600 passenger cars per hour for rural areas as standards. Those numbers are between seventy and seventy-five percent of the basic capacity when the operating speed is assumed to be

[2] Data for each work come from one small subarea.

[3] For example, Choi's[62] regression equation for J is significant only at a 24 per cent significance level.

LOS	70 MPH			60 MPH			50 MPH		
	speed	v/c	MSF	speed	v/c	MSF	speed	v/c	MSF
A	≥ 60	0.35	700	–	–	–	–	–	–
B	≥ 57	0.54	1100	≥ 50	0.49	1000	–	–	–
C	≥ 54	0.77	1550	≥ 47	0.69	1400	≥ 43	0.67	1300
D	≥ 46	0.93	1850	≥ 42	0.84	1700	≥ 40	0.83	1600
E	≥ 30	1.00	2000	≥ 30	1.00	2000	≥ 28	1.00	1900
F	≤ 30			≤ 30			≤ 28		

MSF : Maximum service flow rate per lane under ideal conditions.
Source: HCM 1985 (TRB 1985)[248]

Table 9.1: Levels of Service for Freeways

thirty-five to forty miles per hour. It was this practical capacity that was used in the roadway congestion function developed by the Bureau of Public Roads (BPR 1964)[52].

These notions of basic, possible, and practical capacity were not found in the 1965 HCM[120]. The definition of capacity in the 1965 HCM[120] corresponds to the possible capacity in the 1950 HCM[51]. The concept of *capacity under ideal conditions* replaced basic capacity. Furthermore, the notion of practical capacity was replaced by several *service volumes* that correspond to each *level of service* (LOS). Six service levels, A to F, were then defined. Under ideal roadway conditions, these capacities for each service level can be summarized by passenger car and operating speed (Table 9.2).

the criteria for two-lane highways are listed in Table 9.4.

These capacity concepts become more sophisticated in the 1985 HCM (TRB 1985)[248]. Most notably, *flow rate* becomes the unit of capacity, replacing the old unit of *volume*. By using a flow rate of a peak fifteen-minute period, variations in flow during an hour can be addressed. Furthermore, maximum flow can be analyzed by utilizing the flow rate concept. The concept of six levels of service is retained even though the specific criteria to define these levels of service have been modified. *Service volume* was also changed to *service rate* following the change in the definition of capacity. Level-of-service criteria for freeways with different design speeds are described in Table 9.1. The corresponding criteria for multilane highways are summarized in Table 9.3, while the criteria for two-lane highways are listed in (Table 9.4).

	Freeway	Rural Highway	
		Two-lane	Multilane
LOS A	700 pc/ln/hr	400 pc/hr	600 pc/ln/hr
	$\geq$ 60 mph		60 mph
LOS B	1000 pc/ln/hr	900 pc/hr	1000 pc/ln/hr
	$\geq$ 55 mph		55 mph
LOS C	1500 pc/ln/hr	1400 pc/hr	1500 pc/ln/hr
	50 mph		45 mph
LOS D	1800 pc/ln/hr	1700 pc/hr	1800 pc/ln/hr
	40 mph		35 mph
LOS E	2000 pc/ln/hr	2000 pc/hr	2000 pc/ln/hr
	30 -35 mph		30 mph
LOS F	0-2000 pc/ln/hr	0-2000 pc/hr	0-2000 pc/ln/hr
	< 30 mph		< 30 mph

Source: HCM 1965 (TRB 1965)[120]

Table 9.2: Six Levels of Service Based on the 1965 HCM

LOS	70 MPH			60 MPH			50 MPH		
	speed	v/c	MSF	speed	v/c	MSF	speed	v/c	MSF
A	$\geq$ 57	0.36	700	$\geq$ 50	0.33	650	–	–	–
B	$\geq$ 53	0.54	1100	$\geq$ 48	0.50	1000	$\geq$ 42	0.45	850
C	$\geq$ 50	0.71	1400	$\geq$ 44	0.65	1300	$\geq$ 39	0.60	1150
D	$\geq$ 40	0.87	1750	$\geq$ 40	0.80	1600	$\geq$ 35	0.76	1450
E	$\geq$ 30	1.00	2000	$\geq$ 30	1.00	2000	$\geq$ 28	1.00	1900
F	$\leq$ 30			$\leq$ 30			$\leq$ 28		

MSF : Maximum service flow rate per lane under ideal conditions.

Source: HCM 1985 (TRB 1985)[248]

Table 9.3: Levels of Service for Multilane Highways

LOS	Level Terrain		Rolling Terrain		Mountainous Terrain	
	speed	v/c	speed	v/c	speed	v/c
A	≥ 58	0.15	≥ 57	0.15	≥ 56	0.14
B	≥ 55	0.27	≥ 54	0.26	≥ 54	0.25
C	≥ 52	0.43	≥ 51	0.42	≥ 49	0.39
D	≥ 50	0.64	≥ 49	0.62	≥ 45	0.58
E	≥ 45	1.00	≥ 40	0.97	≥ 35	0.91

Maximum flow rate is 2,800 passenger car per hour in both direction

Source: HCM 1985 (TRB 1985)[248]

Table 9.4: Levels of Service Criteria for Two-lane Highways

As can be seen in these tables, the LOS C capacity is the closest to the 1950 definition of the practical capacity in terms of volume/capacity (V/C) ratio (Table 9.1). LOS C corresponds to a V/C ratio of 0.77 and 0.69 for design speeds of seventy miles per hour and sixty miles per hour, respectively, and is the most probable choice to replicate the practical capacity, for which the V/C ratio was defined in 1950 as 0.7–0.75. For multilane highways, volume between LOS C and LOS D corresponds to the practical capacity based on both the V/C ratio and operating speed (Table 9.3). For two-lane highways (Table 9.4), the 1950 practical capacity corresponds to the volume between the capacities of LOS D and LOS E in terms of V/C ratio. It is important to note that these are only rough comparisons between practical capacity and LOS volume, because practical capacity is a subjective entity, despite the fairly objective guidelines.

Based on this summary, it is assumed that the practical capacity that has been used in the original BPR congestion function most likely corresponds to the LOS C, D, and E of 1985 HCM (TRB 1985)[248], depending on the type of highway. With this conclusion in mind, the next subsection summarizes Korean highway capacity analysis practice.

9.4.2 Highway Capacity Estimation Practice in Korea

Before the advent of the 1985 HCM (TRB 1985)[248], the highway capacity calculation practice of Korea had been based on the 1965 HCM (TRB 1965)[120]. As stated earlier, the Korean highway environment is different from that of the United States. The size of cars, driver characteristics, high heavy-vehicle mix (such as bus and truck), and land use characteristics are unique to Korea.

Cha and Kim (1986)[59] proposed a series of capacities for selected highways in Korea.[4] Accepting the headway and, subsequently, density figures of the 1985 HCM (TRB 1985)[248][5] and applying a shorter car length (five meters) and different design speed, they came up with slightly higher *ideal capacity*[6] figures than those of the 1985 HCM (TRB 1985)[248]. After obtaining the ideal capacity in this way, they applied 1985 HCM procedures based on the national Korean highway environment to calculate the representative capacities for different kinds of highways. Cha and Kim (1986)[59] tested their calculation against actual traffic flow rate observed on various highway segments and reported successful goodness-of-fit between their estimation and actual volume counts. Their results are used as a basis for capacity estimation in this paper pending more comprehensive research on the capacity calculation for the Korean highway environment.

The next section analyzes the procedures involved in developing a link congestion function. A bilevel programming model is introduced to implement the parameter calibration procedures discussed.

9.5 A Bilevel Programming Model for Estimating the Highway Congestion Function

9.5.1 An Overview

The following assumptions and observations are made about developing a link congestion function in Korea.

1. Three different service levels (LOS C, D, and E) of capacity are chosen for the analysis based on the result of the previous section, which suggested that the practical capacity might be closest to the LOS C capacity.

2. In deciding the form of congestion function, the following points are considered:

[4] Highways included are expressway, two-lane expressway, multilane national highways, two-lane national highway, and two-lane unpaved highway.

[5] Vehicle spacing of four car lengths is used in the HCM, producing 67 pc/mi/ln assuming car length is twenty feet. Cha and Kim (1986)[59] used a car length of five meters, producing 50 pc/km/ln.

[6] This corresponds to the capacity under ideal conditions for the 1965 HCM and basic capacity for 1950 HCM.

- Function should be nonnegative and monotonically increasing as the flow increases.
- Function should be able to be decided with the minimum number of parameters.
- Function should replicate the original congestion function (Figure 9.1(b)) well.

Many reported satisfactory performance of the BPR formula (equation 9.6) in the U.S. For example, Branston (1976)[46] reported how the BPR formula (equation 9.6), which has only two parameters, replicates the TRC formula (equation 9.8), which has five parameters. Based on all these points and performance, the BPR-type function is chosen as the base form of congestion function for the Korean environment. The different functional forms, such as the Davidson (1966)[78] function (equation 9.13), can be calibrated using the same basic procedure utilized in this paper.

The usual procedure of the link congestion function development is summarized in four steps. First, a function is devised that can replicate the observed link travel time and volume data. Second, the parameters of the function are calibrated with observed link travel time and volume data, usually based on some statistical means or other mathematical techniques (see, for example, Taylor, 1977)[244]. Next, this function is utilized in an assignment process. Finally, the predicted link volume from the assignment process is tested against the actual link volume count. If the test results are unsatisfactory, the whole process might be repeated.

The alternative procedure used in this paper is different from the previous approach in the data requirements. The observed data of link travel flows at various speeds are not readily available and are very expensive to obtain. Therefore, the alternative process utilizes only the observed link volume as input data. The annual report on link volume count by the Ministry of Construction (MOC) of Korea is utilized. In fact, the method used here actually combines the second, third, and fourth steps mentioned before and can be summarized in two steps. First, a functional form is proposed (in this case, the BPR formula (equation 9.6)). Second, a bilevel programming model is used to calibrate the parameters that can minimize the sum of the least square of the difference between observed link flows and predicted link flows. The advantage of using bilevel programming is that it can specifically restrict the flows to follow the user-optimal principle during the systematic parameter estimation process. Figure 9.3 shows the general steps that will be followed in this paper. The next subsection introduces the bilevel programming model utilized; but first, a brief introduction to the bilevel programming problem is in order.

9.5.2 The Model

Specifically, the model utilized in this paper to implement the alternative approach is as follows:

$$\textbf{P9.2)}\quad \textbf{U9.2)}\quad \min_{\alpha,\beta} \sum_a (f_a^\circ - f_a(\alpha,\beta))^2 \tag{9.16}$$

$$\alpha, \beta \geq 0 \tag{9.17}$$

where f solves

$$\textbf{L9.2)}\quad \min_x \sum_a \int_0^{f_a} c_a(x,\alpha,\beta)dx \tag{9.18}$$

$$\text{s.t.}\quad \sum_p X^{ijp} = T^{ij} \tag{9.19}$$

$$X^{ijp} \geq 0 \tag{9.20}$$

where

$$\begin{aligned} c_a(x,\alpha,\beta) &= t_0[1+\alpha(x/C)^\beta] \\ f_a &= \sum_i \sum_j \sum_p X^{ijp} \delta_a^{ijp} \\ f^0 &: \text{observed link volumes} \\ X^{ijp} &: \text{volume of path } p \text{ connecting zones } i, \text{ and } j \\ T^{ij} &= \text{traffic volume between zone } i \text{ and } j \\ \delta_a^{ijp} &= \begin{cases} 1 & \text{if link } a \text{ is on the path } p \\ 0 & \text{otherwise.} \end{cases} \end{aligned}$$

U9.2) utilizes a least square measurement (equation 9.16) to minimize the sum of the squared differences between the observed link flow and predicted link flow by adjusting α and β of the congestion function. Different values of α, β passed to **L9.2)** from the **U9.2)** cause different link congestion costs and, subsequently, different user-optimal link flow patterns in the **L9.2)**. **L9.2)** is, in fact, a user-optimal assignment programming model. Measurements other than the least square measurement can also be utilized as objective functions in **U9.2)**. One example is an χ^2 measurement ($\frac{(f^\circ - f)^2}{f}$). Maximum likelihood formulation is also possible (for example, see

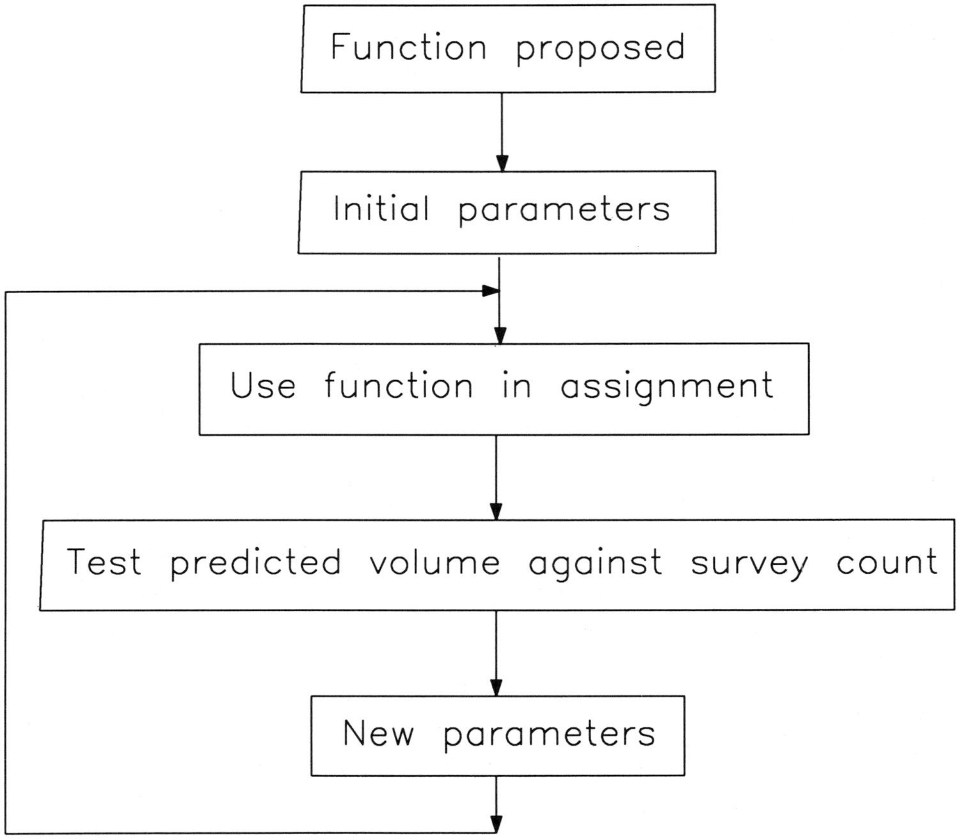

Figure 9.3: Procedures Utilized in Chapter 9

Lee 1986)[162]. For this research, the least square measurement is utilized. Daganzo (1977)[74] reported that the least square measurement performed as well as χ^2 or the maximum likelihood measurement in a small stochastic network problem.

The next section describes a case study of the model. The model is applied to the Korean highway system to calibrate α, β of the BPR function.

9.6 Highway Congestion Function for Korea: An Estimation

The proposed model has a unique structure in the sense that there are only two upper-level variables. Despite the fact that there are only two variables involved, it is beyond the scope of this paper to describe detailed algorithms for solving bilevel programming problems. Detailed descriptions of solution procedures are omitted here, since Suh and Kim (1988)[238] and Suh (1989)[236] report such descriptions in detail.

The lower-level problem L9.2) is solved with the Frank-Wolfe algorithm (LeBlanc et al. 1975)[160]. The network shown in Figure 2.4 and the O/D tables of 1985 listed in Table 9.5 are used for the lower-level problem. There are twenty-five O/D zones, 212 nodes, and 706 links in the network.

Table 9.6 shows the computation results. The computation was done on the Cray XMP/48 at the National Center for Supercomputing Applications (NCSA) at the University of Illinois at Urbana-Champaign. The starting point $\alpha = 0.15$, $\beta = 4$ (BPR formula [52]), and $\alpha = 2.62$, $\beta = 5$ (Steenbrink [233]) are used with LOS E, C, and D capacities in the model. The existence of multiple local optima are identified in the solution.

The parameter values of $\alpha = 2.72$ and $\beta = 6$ are found to be the best fit when LOS D capacity is used with the given network and O/D table. The least square measurement is reduced by about thirty percent using new parameter values over old BPR parameters. Figure 9.4 compares parameter values thus found with those of Steenbrink (1974)[233] and BPR when free flow travel time (T_o) is 1.

9.7 Concluding Remarks

The purpose of this paper has been to calibrate the parameters of the BPR congestion function for the Korean highway system. Toward this end, the paper first identified that LOS C, D, and E service volumes in the 1985 HCM

O/D	1	2	3	4	5	6	7	8	9	10	11	12	13
1	0	21978	7659	9653	22234	693	679	498	865	1485	953	1686	2969
2	24873	4839	925	1346	10387	345	118	77	73	375	64	509	499
3	8219	961	96	2643	464	170	16	16	27	98	9	57	34
4	9985	1609	2412	131	2691	692	378	673	121	69	26	86	93
5	23291	8408	502	2645	4858	114	145	126	74	464	425	1530	2546
6	789	238	114	659	86	114	669	17	557	14	0	0	8
7	907	89	10	322	192	682	241	1192	887	17	9	5	33
8	1105	326	12	585	167	33	1261	70	80	210	21	1034	168
9	1066	59	8	115	122	460	856	52	470	13	2	7	44
10	1577	340	19	61	490	3	19	111	14	136	4057	1057	1587
11	1174	49	9	26	346	0	10	52	2	3965	118	677	2917
12	2233	264	30	487	1186	2	21	1363	8	1619	376	354	659
13	3240	425	261	96	2425	8	30	100	28	1243	2693	680	258
14	1882	364	25	56	403	5	24	10	24	2354	2101	102	576
15	1380	201	18	43	250	10	23	23	43	433	20	60	145
16	758	207	15	44	202	2	13	33	17	810	20	430	209
17	266	36	65	81	44	0	96	348	62	55	0	97	26
18	380	101	6	26	130	6	13	18	634	111	9	14	46
19	2005	772	166	158	1056	17	42	43	71	405	35	121	141
20	433	196	4	22	162	1	11	9	22	212	6	17	95
21	706	233	10	35	254	1	7	14	13	208	29	50	120
22	161	17	50	12	93	0	6	3	2	30	9	11	13
23	1034	132	17	9	148	7	6	10	5	140	20	21	58
24	301	41	3	10	57	1	6	3	6	74	26	23	74
25	335	25	6	7	43	3	7	2	7	65	31	8	32
TTL	88100	41812	12442	19271	48489	3370	4697	4864	4113	14603	11059	8640	13349

O/D	14	15	16	17	18	19	20	21	22	23	24	25	Total
1	1558	1205	807	139	313	1711	326	479	114	882	109	208	79201
2	368	208	196	36	79	746	86	151	26	91	28	27	46470
3	24	28	16	39	6	109	4	14	91	6	2	2	13050
4	51	60	57	95	23	161	19	38	5	9	11	5	19498
5	305	239	279	33	113	1033	116	247	205	108	42	18	47866
6	13	5	1	0	9	15	1	15	0	0	0	0	3324
7	22	21	8	42	16	35	7	25	2	5	2	2	4774
8	44	22	52	409	18	51	30	12	9	6	5	7	5737
9	13	75	8	42	661	60	28	29	2	5	2	7	4206
10	2483	477	837	38	83	390	186	191	28	143	45	37	14411
11	2287	26	21	0	7	37	5	22	9	23	25	11	11819
12	134	54	492	130	9	118	10	44	6	16	21	11	9648
13	537	147	154	22	76	180	73	113	15	58	66	16	12943
14	1334	274	185	9	33	632	50	187	583	2868	247	1439	15763
15	252	902	7035	2880	4656	3200	1299	3163	1828	215	91	84	28255
16	185	6494	288	2062	373	797	582	1004	883	88	17	42	15574
17	12	2442	2120	277	759	187	60	40	14	15	5	5	7110
18	45	3503	432	874	486	2272	2092	861	41	57	13	83	12255
19	511	3413	991	176	2360	0	9344	11433	1950	527	93	459	36290
20	78	1505	700	42	2137	9417	987	3216	145	120	19	58	19616
21	207	3224	1032	40	864	11745	3239	3585	4376	410	108	286	30794
22	681	1629	850	26	37	2339	165	3975	426	161	75	1106	11876
23	2892	227	394	1	43	521	79	381	163	418	6995	3260	16982
24	277	111	20	5	16	171	18	124	78	6051	548	1063	9109
25	1481	133	63	7	20	581	39	352	1233	2828	1360	500	9167
TTL	15791	26422	17039	7422	13200	36509	18842	29699	12231	15108	9930	8737	485740

Table 9.5: The Observed Trip Table for 25 O/D Zone in 1985

Starting Point	Cap.	Solution	F	Iteration	CPU (sec)
	LOS D	$\alpha = 0.45$ $\beta = 7$	8465E10	111	42.00
$\alpha = 0.15$ $\beta = 4$	LOS C	$\alpha = 0.41$ $\beta = 7.21$	7165E10	151	49.73
	LOS E	$\alpha = 1.65$ $\beta = 9$	8685E10	122	34.00
	LOS D	$\alpha = 2.72$ $\beta = 6$	6843E10	108	44.00
$\alpha = 2.62$ $\beta = 5$	LOS C	$\alpha = 2.7$ $\beta = 6$	6943E10	108	46.41
	LOS E	$\alpha = 2.92$ $\beta = 2$	6578E10	8	16.00*

* : This calibration is stopped when α, β become negative in the search process. The value is not a local optimal solution.

Table 9.6: Calibration Results for the Link Congestion Function

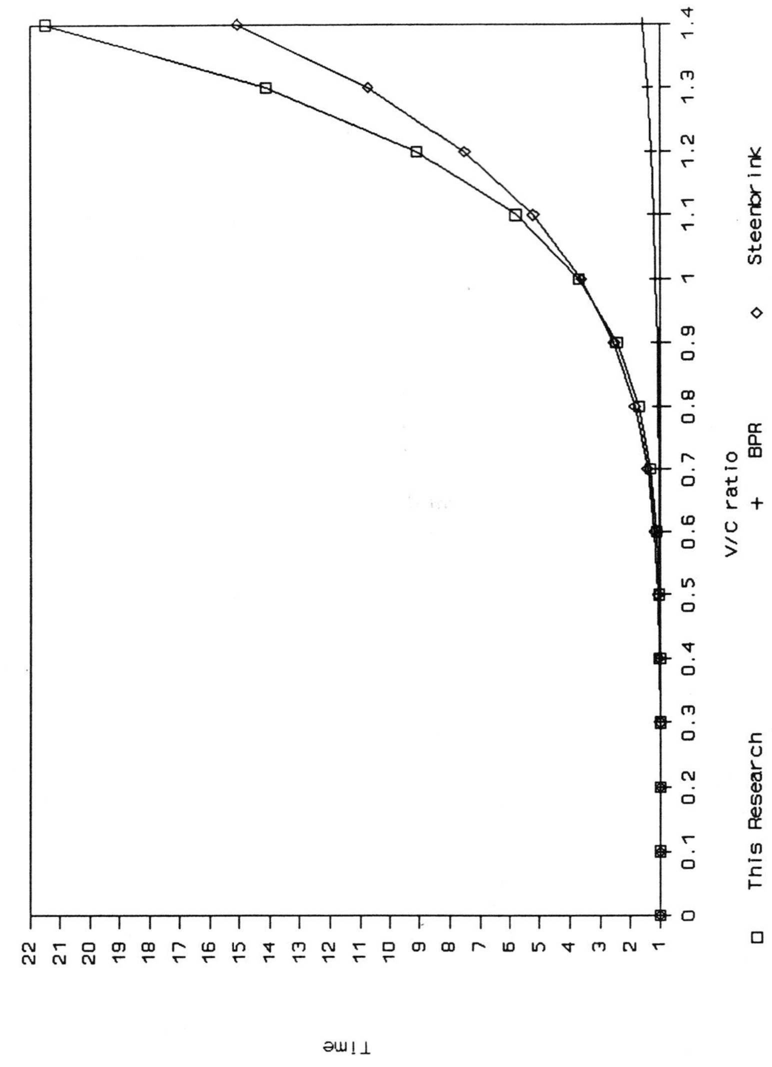

Figure 9.4: Comparison of Congestion Functions with $T_0 = 1$.

correspond to the practical capacity used in the BPR congestion formula, depending on the type of highway.

The link congestion function is calibrated based on a BPR-type formula utilizing an alternative method called a bilevel programming model. The alternative method uses link survey counts instead of link flows at various speeds, which are usually not available. Based on the several computational analyses using different starting points and capacity levels, $\alpha = 2.72$ and $\beta = 6$ are found to be the best fit for Korean highways.

Chapter 10

Model Application II: Evaluating Alternative Highway Investment Strategies

10.1 Introduction

Considering the dominant role of the highway system as a mode of transporting passengers and freight in Korea, it is important to maintain proper highway infrastructure to facilitate national transportation.[1] Therefore, proper decisions on allocating investment capital are also important, considering the large amount invested on highway network upkeep.

In deciding on investments on highway infrastructure, it has been argued thus far that a nonlinear bilevel model of the equilibrium network design problem (ENDP), that can accommodate both private and public sectors is superior in concept to the system optimization network design problem or Cournot-Nash equilibrium model. It has also been stated that different countries might need different link congestion functions and investment functions because distinct social- and transportation-related characteristics exist for each. These are the basic rationales for applying a nonlinear bilevel model

[1]This chapter has been written with Sunduck Suh and draws heavily upon his doctoral dissertation, *Implementation and Evaluation of Nonlinear Bilevel Programming Model of Equilibrium Network Design Problem*, University of Illinois at Urbana-Champaign, Urbana, IL. January, 1989.

of ENDP with an explicit budget constraint to the Korean interregional highway network.

10.2 Data

The Korean regional highway networks are aggregated into the network of fifteen O/D zones, sixty-three nodes, and 190 links. Table 10.1 shows the 1985 vehicle O/D tables for fifteen O/D zonal systems that are compiled from the O/D surveys of KRIHS (1985)[152] and RDRI/KIST (1980)[139].

Either all the links in the network are considered for improvement or a set of links for improvement are developed for network design problem application. With a linear model of the network design problem, all the links are usually considered as alternatives while a set of links are selected a priori with a nonlinear network design problem. Otherwise the model becomes computationally complex.

There are many different ways of identifying bottleneck links and generating alternative links for improvement. In this chapter, the disutility concept method of Kim, Park, and Rho (1985) [141] is used. Their approach is to rank potential links based on disutility, which is the area under the total cost curve divided by the lengths of the links. The candidate links for improvement thus identified a priori are shown in Figure 2.5. Notice that the link numbers do not necessarily represent priorities for investment.

10.3 Numerical Analysis

The bilevel programming problem of the ENDP developed in the previous chapter, P8.5), is solved in this section with the network and alternatives shown in Figure 2.5, the O/D table shown in Table 10.1, the link congestion function of $T = T_o(1 + 2.72(V/C)^6)$, and the investment function developed for the Korean highway environment (Suh 1989)[236].

$$
\begin{aligned}
\text{Total Cost} &= 2.00X^{0.57} \text{ for flat terrain} \\
\text{Total Cost} &= 3.09X^{0.57} \text{ for hilly terrain} \\
\text{Total Cost} &= 6.61X^{0.57} \text{ for mountainous terrain}
\end{aligned}
$$

The lower-level problem (L8.4) is solved with the Frank-Wolfe algorithm[99], the upper-level problem (U8.3) by VF03 of Harwell Library[3].

O/D	1	2	3	4	5	6	7	8
1	69450	198	523	194	246	362	1117	4092
2	2287	1440	36	25	89	148	7	121
3	5696	9	12359	4904	1907	1003	273	6690
4	2270	14	4556	1334	459	369	3115	587
5	2580	36	1773	436	14719	10045	411	410
6	4525	33	1085	796	10865	52966	1277	421
7	1530	11	304	3169	752	1294	14012	146
8	5368	71	5881	581	374	459	135	596
9	227	0	50	681	2479	6478	237	16
10	488	557	135	45	3936	5224	70	65
11	367	11	152	12	4562	287	19	374
12	32201	115	2420	305	518	1396	150	2672
13	14005	769	181	51	117	218	20	765
14	1007	1440	31	22	30	67	7	1226
15	366	3	104	1481	195	972	4188	34
TTL	142367	6493	34302	15791	43462	85050	25038	18214

O/D	9	10	11	12	13	14	15	total
1	231	397	214	33084	13642	312	237	138531
2	2	670	42	239	784	2462	7	8360
3	43	100	168	2022	574	51	59	35858
4	583	33	9	403	56	24	1439	15749
5	2710	5030	4941	453	87	36	126	43792
6	6471	5361	257	1471	214	60	802	86607
7	241	59	6	205	19	11	4324	26083
8	24	95	431	2592	681	1292	23	18602
9	426	37	26	93	12	6	1106	11874
10	41	486	874	130	26	13	83	12173
11	14	759	277	44	81	96	5	7059
12	205	113	33	4858	2645	145	18	47793
13	5	23	95	2691	131	378	5	19454
14	2	16	42	192	322	241	2	4645
15	1233	20	7	43	7	7	500	9161
TTL	12231	13200	7422	48519	19281	5634	8737	485741

Table 10.1: The Observed Trip Table for 15 O/D zone in 1985

The linear system of equations, which is necessary to get the first order derivatives of link flows with respect to link improvements, is solved by PCGPAK[218].

10.3.1 Evaluation of the Model

The objective of this subsection is to assess the model by comparing the model solution with the known link improvement. By this comparison, the model as well as the the known link improvement can be validated based on the solution of the model.

Two series of computer runs are initially carried out when six and twenty-two directional links are considered for improvement. For each series of runs, two different budget levels (50 and 100 billion Won) are imposed. The super-computer Cray XMP/48 is utilized at the National Center for Supercomputing Applications (NCSA) at the University of Illinois at Urbana-Champaign.

Table 10.2 shows the detailed solution process for the problem when six directional links are considered for improvements with a 50 billion Won budget. These six directional links correspond to three undirectional links, e.g., Links 1, 2, and 3 (Figure 2.5).

The program requires 1,120 seconds of CPU time on the Cray XMP/48. Results shown after nine iterations indicate that three undirectional links (Links 1, 2, and 3) can be further expanded in their capacity. This can be substantiated by the fact that the gradients of the total system cost with respect to the capacity expansion on these links are still negative. This means that the objective function, total system cost, can be further reduced with the addition of more capacity on these links. Furthermore, the budget is almost depleted.

Therefore, another run of the model is carried out with a 100 billion Won budget for the same three undirectional links. It required 1,850 seconds of CPU time on the Cray XMP/48. The summary of computation is shown in Table 10.3.

As is depicted in Table 10.3, the gradients of the objective function with respect to Links 1 and 3 are zero. However, the gradients of the objective function with respect to Link 2 is not zero even at the fifteenth iteration, as shown in rows $\partial F/\partial x_3$ and $\partial F/\partial x_4$. The program terminated after fifteen iterations because the total system cost reduction that can be accomplished by the expansion of Link 2 is less than the convergence criteria. The convergence criteria used is 0.02 percent, which means that the program stops when the changes of the objective function value between iterations become

	Iteration					
	0	1	2	3	4	5
Link 1 (s)	0	19	215	312	571	778
(n)	0	22	233	336	605	821
Link 2 (s)	0	0	3	4	8	11
(n)	0	0	4	5	10	14
Link 3 (s)	0	53	603	847	1453	1922
(n)	0	82	906	1248	2099	2760
F	2364210	2250450	1691430	1579940	1409530	1354150
$\partial F/\partial x_1$	-19.3	-19.6	-14.6	-15.4	-9.4	-6.1
$\partial F/\partial x_2$	-21.6	-21.1	-15.3	-13.8	-9.4	-6.1
$\partial F/\partial x_3$	-0.2	-0.2	-0.2	-0.3	-0.3	-0.3
$\partial F/\partial x_4$	-0.3	-0.3	-0.3	-0.3	-0.4	-0.5
$\partial F/\partial x_5$	-53.3	-54.9	-21.6	-14.3	-5.0	-2.3
$\partial F/\partial x_6$	-82.3	-82.2	-16.6	-10.3	-3.5	-1.1
Budget Left (million Won)	50000	45697	33008	29474	22169	17375

	Iteration				Total
	6	7	8	9	
Link 1 (s)	991	1276	1625	1740	
(n)	1042	1339	1703	1822	3561
Link 2 (s)	14	20	26	29	
(n)	19	26	35	39	68
Link 3 (s)	2397	3033	3811	4064	
(n)	3429	4323	5417	5774	9838
F	1321540	1295040	1282700	1277940	
$\partial F/\partial x_1$	-3.8	-2.4	-1.2	-0.9	
$\partial F/\partial x_2$	-3.7	-2.3	-1.1	-0.9	
$\partial F/\partial x_3$	-0.4	-0.5	-0.5	-0.5	
$\partial F/\partial x_4$	-0.5	-0.6	-0.7	-0.7	
$\partial F/\partial x_5$	-1.2	-0.5	-0.2	-0.1	
$\partial F/\partial x_6$	-0.5	-0.2	-0.1	-0.0	
Budget Left (million Won)	13003	7703	1836	25	

Table 10.2: Computation Detail for Six Alternatives with a 50 Billion Won Budget (Unit: PCU's)

	Iteration							
	0	1	2	3	4	5	6	7
Link 1 (s)	0	19	215	312	571	778	991	1276
(n)	0	22	233	336	605	821	1042	1339
Link 2 (s)	0	0	3	4	8	11	14	27
(n)	0	0	4	5	10	14	19	26
Link 3 (s)	0	53	603	847	1453	1922	2397	3033
(n)	0	82	906	1248	2099	2760	3429	4323
F / 10	236421	225045	169143	157994	140953	135415	132154	129504
$\partial F/\partial x_1$	-19.3	-19.6	-14.6	-15.4	-9.4	-6.1	- 3.8	-2.4
$\partial F/\partial x_2$	-21.6	-21.1	-15.3	-13.8	-9.4	-6.1	- 2.4	-1.11
$\partial F/\partial x_3$	-0.2	-0.2	-0.2	-0.3	-0.3	-0.3	- 0.5	-0.5
$\partial F/\partial x_4$	-0.3	-0.3	-0.3	-0.3	-0.4	-0.5	- 0.7	-0.7
$\partial F/\partial x_5$	-53.3	-54.9	-21.6	-14.3	-5.0	-2.3	- 0.5	-0.2
$\partial F/\partial x_6$	-82.3	-82.2	-16.6	-10.3	-3.5	-1.1	- 0.5	-0.2
Budget Left (million Won)	100000	95697	83008	79474	72169	67375	63003	57703

	Iteration							
	8	9	10	11	12	13	14	15
Link 1 (s)	1625	1907	2222	2657	3105	3571	3571	4552
(n)	1702	1996	2324	2776	3243	3727	3728	4757
Link 2 (s)	27	33	40	51	63	76	76	103
(n)	35	43	53	67	83	99	98	135
Link 3 (s)	3811	4439	5139	6104	7101	8137	8139	10317
(n)	5417	6300	7285	8643	10046	11503	11505	14570
F / 10	128270	127954	127131	126415	126175	125597	125287	125037
$\partial F/\partial x_1$	-1.2	-0.7	-0.4	-0.2	-0.1	-0.1	- 0.1	-0.0
$\partial F/\partial x_2$	-1.1	-0.6	-0.3	-0.2	-0.1	-0.0	-0.1	-0.0
$\partial F/\partial x_3$	-0.5	-0.5	-0.5	-0.5	-0.5	-0.5	-0.5	-0.5
$\partial F/\partial x_4$	-0.7	-0.6	-0.7	-0.7	-0.6	-0.6	-0.6	-0.6
$\partial F/\partial x_5$	-0.2	-0.1	-0.0	-0.0	-0.0	-0.0	-0.0	-0.0
$\partial F/\partial x_6$	-0.0	-0.0	-0.0	-0.0	-0.0	-0.0	- 0.0	-0.0
Budget Left (million Won)	51830	47455	42881	36988	31311	25765	25757	15007

Table 10.3: Computation Detail for Six Alternatives with a 100 Billion Won Budget(Unit: PCU's)

Budget	Link					
	1	2	3	4	5	6
50 billion Won	988	0	2909	272	103	5
100 billion Won	3049	16	8363	2251	1791	660

Budget	Link					Objective Function
	7	8	9	10	11	
50 billion Won	510	1555	944	68	93	1305190
100 billion Won	1379	4365	2993	251	73	1005180

Table 10.4: Capacity Improvement for 22 Link Alternatives (Unit: PCU's)

less than 0.02 percent.

With a 100 billion Won budget, it is suggested that Link 3 should be improved to accommodate an additional 24,887 vehicles per day, which is roughly equivalent to the capacity of a four-lane highway. The capacity expansion on Link 1 (9,309 vehicles per day) roughly corresponds to the addition of a two-lane highway. The capacity improvement on Link 2 is inconsequential (239 vehicles per day).

Table 10.4 summarizes the solution of the model when twenty-two links are a priori selected and considered for improvement with 50 billion Won and 100 billion Won. The CPU time required was 900 seconds for the former and 1,420 seconds for the latter, with a convergence criteria of 0.04 percent. Notice that both budget levels are very small for links to have meaningful improvement for every link of eleven. The trend identified before, however, is also present in Table 10.4. One group of links received substantial improvements while the other group received only marginal improvements.

A finding to note is that the CPU time does not have a striking relationship with the number of alternatives considered. To analyze this problem, the time spent on the solution of the upper-level problem, the lower-level problem, and derivative calculation are recorded. For a typical problem, about ninety-one percent of the total CPU time is spent on solving for derivatives, five percent on solving the lower-level problem, and four percent on solving the upper-level problem. In fact, the solutions required only about sixty to seventy Frank-Wolfe iterations. With a given network and O/D table, the CPU time seems to have a more direct relationship with the number of iterations in the problems solved. Therefore, the efficiency of calculating the derivatives is the crucial factor in the overall efficiency of the algorithm.

This section demonstrated the applicability of the nonlinear bilevel programming model of the equilibrium continuous network design problem by applying the model to a real-life transportation planning situation. Using 1985 data, the model successfully confirmed the addition of a four-lane expressway connecting Nodes 16 and 18 (dotted line in Figure 2.4), which took place in 1987.

10.3.2 Policy Implication of the Model Results

In contrast to the model application in the previous subsection, where the outcome of the model is compared with the link addition that had already taken place, this subsection describes an application of the model to identify the optimal level of capacity expansion of links. The main interest of this subsection is the investment plan of a four-lane expressway connecting Nodes 56, 42, 59, 61, and 23 (Figure 2.5). This investment plan is already established by the Korean government and the first segment, connecting Nodes 23 and 61, is expected to be initiated before 1990.[2]

In this application, the model is first applied to the new network, which includes a new four-lane expressway connecting Nodes 16 and 18 (the dotted line in Figure 2.5. The 1985 O/D tables are still utilized. The same twenty-two directional links, except Link 1, which is replaced by Link 13, are considered for improvements for this model application. Two budget levels, of 50 and 100 billion Won, are imposed. Table 10.5 shows the summary of the solution for the two model applications.

As can be identified from Table 10.5, Link 12 (node 61 and 23), which is the first segment to be constructed among the expansion plan, is recommended for significant capacity improvement. The capacity improvement level, however, is not large enough to warrant a four-lane expressway. The capacity level corresponds to the capacity improvement from a two-lane high-

[2]The Daily Dong-A, July 30, 1987

Budget	Link					
	12	2	3	4	5	6
50 billion Won	822	2335	3	0	503	86
100 billion Won	2739	7147	62	196	1733	331

Budget	Link					Objective Function
	7	8	9	10	11	
50 billion Won	363	1736	1805	337	23	1098610
100 billion Won	1250	5320	5680	1111	118	1016470

Table 10.5: Capacity Improvement for 22 Link Alternatives for an Expanded Network

way to a two-lane expressway.[3] Observe that Link 3, one of the busiest link connecting Suweon and Cheonan (nodes 17 and 18), has already been improved to a four-lane highway, and is recommended not to increase its capacity. It can be concluded that Link 3, with the opening of the four-lane expressway connecting Nodes 16 and 18, has enough capacity for the traffic for the time being connecting Seoul and Daejon.

To analyze the problem in more detail, two runs of the model are carried out using only eight directional alternatives (Links 12, 13, 14, and 15 in Figure 2.5 with budget of 50 billion Won. The O/D table of 1985 is still used. Table 10.6 shows the detailed solution of the model with eight alternatives for a 50 billion Won budget.

Again, the capacity improvements identified from the model are not large enough for a four-lane expressway. When the model is applied again with the 100 billion Won budget, no results are obtained that are significantly different from the application with the 50 billion Won budget, except that a large unused budget remains.

[3]One can get 3,500 vehicles per day improvement by expanding a two-lane highway to a two-lane expressway when the terrains are considered hilly.

	Iteration					
	0	1	2	3	4	5
Link 12 (s)	0	12	166	248	354	480
(n)	0	10	116	175	250	340
Link 13 (s)	0	0	0	0	0	0
(n)	0	0	0	0	0	0
Link 14 (s)	0	14	145	214	302	408
(n)	0	8	117	175	248	336
Link 15 (s)	0	0	4	6	8	11
(n)	0	0	1	2	3	5
F	1505180	1496270	1441240	1423730	1416350	1409370
$\partial F/\partial x_1$	-12.4	-12.9	-4.6	-2.7	-1.6	-0.8
$\partial F/\partial x_2$	-10.5	-8.7	-4.1	-2.5	-1.4	-0.8
$\partial F/\partial x_3$	-0.0	-0.0	-0.0	-0.0	-0.0	-0.0
$\partial F/\partial x_4$	-0.0	-0.0	-0.0	-0.0	-0.0	-0.0
$\partial F/\partial x_5$	-13.6	-10.5	-3.1	-1.5	-0.7	-0.3
$\partial F/\partial x_6$	-8.2	-9.3	-2.8	-1.5	-0.8	-0.3
$\partial F/\partial x_7$	-0.3	-0.3	-0.2	-0.1	-0.1	-0.1
$\partial F/\partial x_8$	-0.1	-0.1	-0.1	-0.1	-0.1	-0.1
Budget Left (million Won)	50000	47031	37496	34255	30733	27038

	Iteration				Total
	6	7	8	9	
Link 12 (s)	603	743	879	1098	
(n)	428	528	625	782	1880
Link 13 (s)	0	0	0	0	
(n)	0	0	0	0	0
Link 14 (s)	510	626	739	921	
(n)	422	518	613	765	1686
Link 15 (s)	15	18	22	28	
(n)	6	8	10	13	41
F	1399080	1392860	1388780	1380870	
$\partial F/\partial x_1$	-0.4	-0.2	-0.1	-0.0	
$\partial F/\partial x_2$	-0.4	-0.3	-0.2	-0.0	
$\partial F/\partial x_3$	-0.0	-0.0	-0.0	-0.0	
$\partial F/\partial x_4$	-0.0	-0.0	-0.0	-0.0	
$\partial F/\partial x_5$	-0.1	-0.1	-0.0	-0.0	
$\partial F/\partial x_6$	-0.2	-0.1	-0.1	-0.0	
$\partial F/\partial x_7$	-0.1	-0.1	-0.1	-0.1	
$\partial F/\partial x_8$	-0.1	-0.1	-0.1	-0.1	
Budget Left (million Won)	23837	20522	17535	13118	

Table 10.6: Computation Detail for Eight Alternatives with a 50 Billion Won Budget

What if the model applied with the current O/D table represents 1988 traffic demand? Is the same improvement pattern going to be identified? To answer these questions, the model is applied with an arbitrary O/D trip table that represents twenty percent increases from the 1985 O/D trip table. Table 10.7 represents the summary of the model solution for the increased traffic demand with a 100 billion Won budget. As can be identified from Table 10.7, the capacity improvements increase with increased traffic demand. When the expressway connecting Nodes 23 and 61 opens in the mid-1990s, the traffic growth might be large enough for a four-lane expressway. Currently, the traffic demand level using the alternative links is not large enough for a four-lane expressway, as shown by the model application. The capacity level, however, is large enough for expanding a two-lane highway to a two-lane expressway.

10.4 Concluding Remark

Public sector's inherent involvement with transportation necessities many traditional models to be recast into a bilevel programming problem. The removal of the fixed demand assumption can further enhance the applicability of the bilevel network design problem to national strategic transportation planning, as is shown in the chapter that follows.

In terms of solving such bilevel network design problems, the more efficient the method to calculate the derivative information of the lower-level problem, the less the computational burden to solve the overall problem. Thus, further research to devise a more efficient procedure for sensitivity analysis is in order when considering the fact that about ninety percent of computation time is spend on derivative calculation. Currently, the algorithm deals with path flow variables in calculating derivatives. The employment of path flow in the derivative calculation, even though the algorithm deals only with the shortest path between O/D pair, can be a potential problem when large numbers of O/D pairs are involved.

The refinement of program codes also deserves much attention. More specifically, attention should be given to developing more efficient methods of storing and manipulating sparse matrices in derivative calculations. Some specific schemes also deserve high priority, such as *vectorization* of the codes and the adoption of *parallel processing* for calculations to utilize the new hardware of the new generation of supercomputers.

Because of the computational complexity inherent to both the network

Links	Iteration					
	0	1	2	3	4	5
12 south	0	53	326	509	699	832
12 north	0	35	216	340	472	567
13 south	0	0	0	0	0	0
13 north	0	0	0	0	0	0
14 south	0	39	169	293	392	462
14 north	0	50	221	333	447	527
15 south	0	2	8	13	18	22
15 north	0	1	4	7	11	14
F	3330050	3220840	2973640	2860660	2817030	2780880
$\partial F/\partial x_1$	-52.7	-50.2	-21.5	-10.7	-3.9	-2.7
$\partial F/\partial x_2$	-35.3	-33.1	-15.8	-10.1	-4.7	-3.3
$\partial F/\partial x_3$	-0.0	-0.0	-0.0	-0.0	-0.0	-0.0
$\partial F/\partial x_4$	-0.0	-0.0	-0.0	-0.0	-0.0	-0.0
$\partial F/\partial x_5$	-39.1	-27.8	-6.3	-2.8	-1.2	-0.7
$\partial F/\partial x_6$	-49.7	-29.3	-10.6	-4.0	-1.8	-1.1
$\partial F/\partial x_7$	-1.8	-1.1	-0.8	-0.5	-0.5	-0.4
$\partial F/\partial x_8$	-0.7	-0.6	-0.7	-0.7	-0.6	-0.5
Budget Left (M. Won)	100000	93413	82108	76932	72297	69329

Links	Iteration					Total
	6	7	8	9	10	veh/day
12 south	1113	1304	1533	1809	2243	
12 north	766	903	1066	1263	1574	3817
13 south	0	1	1	2	3	
13 north	0	0	0	0	0	3
14 south	610	711	832	977	1207	
14 north	696	812	950	1117	1380	2588
15 south	31	38	45	55	71	
15 north	21	26	33	41	54	125
F	2742640	2730540	2716080	2699680	2677620	
$\partial F/\partial x_1$	-0.9	-0.5	-0.3	-0.1	-0.0	
$\partial F/\partial x_2$	-1.5	-0.8	-0.5	-0.3	-0.1	
$\partial F/\partial x_3$	-0.1	-0.1	-0.1	-0.3	-0.2	
$\partial F/\partial x_4$	-0.0	-0.0	-0.0	-0.0	-0.0	
$\partial F/\partial x_5$	-0.4	-0.2	-0.1	-0.1	-0.0	
$\partial F/\partial x_6$	-0.4	-0.3	-0.1	-0.1	-0.1	
$\partial F/\partial x_7$	-0.4	-0.4	-0.3	-0.3	-0.2	
$\partial F/\partial x_8$	-0.4	-0.4	-0.3	-0.3	-0.2	
Budget Left (M. Won)	63630	60065	56054	51526	44903	

Note : See Table 2 for explanation of symbols

Table 10.7: Computation Detail for Eight Alternatives with Expanded Demand and a 100 billion Won Budget

design problem and the bilevel programming problem, current hardware is inadequate for actual use in solving very large-scale problems. Therefore, development of heuristic approaches to solving very large-scale problems also deserves much attention.

Part V

ADVANCED TRANSPORT AND SPATIAL PLANNING MODELS: FUTURE PROSPECTS

Chapter 11

Integrated National Transport Planning Model for Mixed Economic Systems: A Bilevel Programming Approach

11.1 Introduction

As was described previously, at least two parties are involved in the decision-making processes in mixed economic systems: the public sector, which constructs new infrastructure systems, improves their capacities, and regulates services and prices; and users, who choose locations of residence and production, modes of transportation, and shipment routes.[1] While many have acknowledged the existence of this type of multilevel decision-making process as an important planning issue, no one has explicitly incorporated it into existing development planning models. Failure to explicitly include the public-private interaction in the model in Korea could result in inconsistent public policies, as described earlier. For example, policies designed to reduce Seoul's dominance within Korea may affect Korea's international competitiveness (Ko and Hewings 1986)[144]. Consideration of development/investment pol-

[1] This chapter, in its first form, was written with Sunduck Suh, and was published in *The Annals of Regional Science*, 1988, 20SPED: 65–80.

icy cannot be divorced from the centripetal forces of private sector economic activities. As the world system of markets becomes more integrated, the ability of the government to pursue development policies without much regard for the private sector will have to be circumscribed.

Bilevel programming models can explicitly analyze two different, conflicting objectives involved in the decision-making processes. Bilevel programming approaches have a conceptual advantage over conventional single-level programming models, because they can explicitly simulate the interaction between different hierarchies of the decision process, such as public-private interaction.

11.2 The Model

A bilevel programming model is proposed for Korea in this section. For the sake of simplicity, transportation is assumed to be the only infrastructure provided by the public sector. The model also includes decision variables by the public sector. Thus, the public sector will make improvements (continuous) or additions (discrete) to an existing transportation network (the upper-level problem), whereas the lower-level problem solves shipment volume on every transportation link that is the result of each individual choice as to location, amount, and type of production, and choice of shipment modes and routes (a user-optimal equilibrium transportation problem). The proposed model is as follows:

$$P11.1)U11.1) \quad \max_{E,y} U = \sum_i \sum_r e_r^i E_r^i - \sum_a \int_0^{f_a} M_a(x(y), y_a)dx \tag{11.1}$$

$$\text{s.t.} \quad \sum_i \sum_r E_r^i \geq E_r \quad \forall \ r \tag{11.2}$$

$$\sum_a G_a(y_a) \leq B \tag{11.3}$$

$$E_r^i, y_a \geq 0 \tag{11.4}$$

where x solves

$$L11.1) \quad \min_x \quad L = \sum_a \int_0^{fa} C_a(x)dx + \sum_i \sum_r p_r^i X_r^i \tag{11.5}$$

$$\text{s.t.} \quad \sum_p X_r^{ijp} = X_r^{ij} \quad \forall \ i, j, r \tag{11.6}$$

$$\sum_j X_r^{ij} + X_r^i = \sum_j X_r^{ij} + \sum_q a_{rq} X_q^i + E_r^i \forall \quad i, r \tag{11.7}$$

$$f_a = \sum_i \sum_j \sum_p \delta_a^{ijp} \sum_r X_r^{ijp} \tag{11.8}$$

$$X_r^{ijp}, X_r^i, X_r^{ij} \geq 0 \tag{11.9}$$

where the exogenous variables are:

e_r^i = FOB price of commodity r at port i,
$M_a(x(y), y_a)$ = marginal shipment cost function for link a,
E_r = total export of commodity r,
$G_a(y_a)$ = investment function for link a,
B = total budget for transportation improvement,
$c_a(x)$ = operating cost function for link a,
p_r^i = unit production cost for commodity r at zone i,
a_{rq} = amount of input q required per unit of output r,

$$\delta_a^{ijp} = \begin{cases} 1, & \text{if link } a \text{ is on the path joining } i \text{ to } j, \\ 0, & \text{otherwise} \end{cases}$$

and the endogenous variables are

E_r^i = export amounts of commodity r from export zone i,
y_a = capacity increase in shipment flow on link a,
X_r^i = output of commodity r at zone i,
X_r^{ij} = units of r shipped from zone i to zone j,
X_r^{ijp} = flow of commodity r on a path p connecting zone i to j.

Korea has been implementing export targets with its series of Five-Year Development Plans. Once export targets are given, the upper-level problem ($U11.1$) is to maximize total export revenues ($\sum_i \sum_r d_r^i E_r^i$) by subtracting from total costs transportation ($\sum_a \int_0^{f_a} M_a(x(y), y_a)dx$) and construction costs for additional networks ($\sum_a G_a(y_a)$) that are necessary to transport produced goods and services for the society as a whole. The lower-level problem (L11.1) is to minimize the sum of the individual's average cost of shipment ($\sum_a \int_0^{f_a} C_a(x)dx$) and the sum of the individual producer's production costs ($\sum_i \sum_r p_r^i X_r^i$) subject to constraints on the conservation of shipment flows (Equation 11.7). The upper-level problem, then, is a system optimization problem, which the public sector strives to enhance by maximizing export revenues and minimizing costs for providing public services. The lower-level problem is a user-optimization problem representing the private sector's maximum profits/minimum costs behavior. The model shown above is typically *nonconvex* and is not directly amenable to the solution by the available nonlinear programming solution algorithms. Some aspects that can shed light on developing an efficient solution algorithm for the model are described below.

The optimality conditions for the lower-level problem, L, are as follows:

$$\begin{aligned} H &= \sum_a \int_0^{f_a} C_a(x)dx + \sum_i \sum_r p_r^i X_r^i + \sum_i \sum_j \sum_r c_r^{ij}(X_r^{ij} - \sum_p X_r^{ijp}) \\ &+ \sum_i \sum_r \gamma_r^i (\sum_j X_r^{ij} + \sum_q a^{rq} X_q^i + E_r^i - \sum_j X_r^{ij} - X_r^i) \\ &+ \sum_i \sum_j \sum_p \sum_r \theta_r^{ijp}(-X_r^{ijp}) + \sum_i \sum_r \alpha_r^i(-X_r^i) \\ &+ \sum_i \sum_j \sum_r \lambda_r^{ij}(-X_r^{ij}) \end{aligned} \tag{11.10}$$

$$\frac{\partial H}{\partial X_r^i} = p_r^i + \sum_q \gamma_q^i a_{rq} - \gamma_r^i - \alpha_r^i = 0 \tag{11.11}$$

$$\frac{\partial H}{\partial X_r^{ij}} = c_r^{ij} + \gamma_r^i - \gamma_r^j - \lambda_r^{ij} = 0 \tag{11.12}$$

$$\frac{\partial H}{\partial X_r^{ijp}} = \sum_a c_a(f_a)\delta_a^{ijp} - c_r^{ij} - \theta_r^{ijp} = 0 \tag{11.13}$$

Where $c_r^{ij}, \gamma_r^i, \theta_r^{ijp}, \alpha_r^i$, and λ_r^{ij} are Lagrange multipliers associated with Equations 11.6, 11.7, and all nonnegative variables in Equation 11.9, respectively. Some interpretations on the optimality conditions are as follows: If $X_r^{ijp} > 0$, then $\theta_r^{ijp} = 0$, which implies in Equation 11.12 that a user-equilibrium condition ($\sum_a c_a(f_a)\delta_r^{ijp} = c_r^{ij}$) is met at optimum. In other words, no unused path has lower cost than the used path between i and j, and all used paths have equal cost, i.e., c_r^{ij} (Wardrop 1952)[252].

A spatial price equilibrium condition ($c_r^{ij} + p_r^i \leq \gamma_r^j$) can also be derived from Equations (11.11) and (11.12) when $X_r^{ij} > 0$. This condition states that, in order to induce positive flows of commodity r from i to j, the sum of transportation costs of shipping r from i to $j(c_r^{ij})$ and the production cost of r at zone $i(p_r^i)$ should be less than or equal to the opportunity cost of producing a unit of commodity r at zone $j(\gamma_r^j)$.

The proposed model P11.1 can also be stated in a more general form, including variational inequality (VI), as follows:

$$\min_{y \in Y} \quad U(x, y)$$

$$\text{s.t.} \qquad (x - z)H(x, y) \leq 0 \text{ for } z \in X(y)$$

$$x \in X(y)$$

where $Y(.)$ is the set of feasible regions for U, and $X(.)$ is the set of feasible region for L.

A few scholars have suggested alternative algorithms for solving bilevel programming problems (see chapter 8 for a further survey of available solution algorithms). None of those solution algorithms, however, have been tested using a large-scale real world problem such as the one shown in this chapter. A solution algorithm that is currently being tested by the author is a descent-type algorithm.

The algorithm is, in essence, an iterative one that works as follows: (1) obtain feasible decision variables for (U); (2) solve VI constraint, which is equivalent to solving (L); (3) compute derivatives with respect to decision variables for (U); and (4) determine the next values for decision variables in (U2) according to the Quasi-Newton update method, using the gradient information obtained in the third step.

11.3 Concluding Remarks

Korea made remarkable economic progress in past decades by the successful implementation of a series of FYDPs that began in 1962. GNP tripled in real terms and exports grew more than 400 times, from \$52 million in 1962 to \$24.5 billion in 1984. No doubt these are the direct results of the "export-oriented" policy that the government has pursued. The export-oriented policy has often been accompanied by investment policy that has favored improving infrastructures in major cities, particularly in Seoul. It is, in fact, the efficiency that the government enhanced that facilitates economies of scale. These scale economies, in turn, often necessitate the agglomeration of socio-economic activities. Decentralization of socio-economic activities from Seoul will not be feasible unless the government implements its investment policy based on the optimum results obtained by the interaction between public and private sectors. Such a model has been introduced here. The model has not been implemented using the real world data yet and no efficient solution algorithm has been developed. Based on the initial development of a solution algorithm described in chapter 8, however, it may not be long before the model can be implemented.

Chapter 12

Future Research Agenda

12.1 Introduction

All forms of transportation influence our lives and give us cause for concern for our environment, health, and safety. In fact, transportation is intimately interwoven with the daily lives of individuals and organizations in our society; thus, it is easy to overlook the significance of transportation until it fails us in some way (Schofer and Boyce 1985)[220].

The amount of investment in transportation infrastructure in any country, particularly in developed countries, is immense. The impact of transportation on efficiency, productivity, and competitiveness of a nation's economy is also tremendous. In the United States, the transportation industry, including service delivery and manufacturing, contributes more than 20% of the Gross National Product (GNP).

In light of these and other indications that suggest the critical importance of transportation to the quality of life, this chapter attempts to assess past and current developments in transportation, particularly in regard to its spatial interaction and telecommunication, in order to better identify future research agenda. Discussions on future research opportunities will be focused on the following two categories:

1. those that are old and continuing issues that we will still have to face continuously as challenging research opportunities;

2. those new issues that are brought about particularly by the advance of the telecommunication technologies and will provide us with new challenging research opportunities.

Toward identifying research agenda in these two categories, the chapter is organized as follows: In the section that follows, changes in the socio-economic environment in relation to transportation are discussed. In Section 12.3, past and current research in transportation systems from the supply side viewpoint are assessed and future research opportunities are discussed. In Section 12.4, research on transportation and spatial interaction from the demand side is reviewed and future research opportunities are identified. In Section 12.5, research on transportation and telecommunication is analyzed and future research directions are suggested in the final

12.2 The Changing Socio-Economic Environment and Transportation

Present urban and regional systems are complex. Spatial analysis models cannot produce useful recommendations to decision makers unless they are founded on an understanding of the continuously changing nature of structures and the behavior of the actors involved in decision making who affect the functions and structure of an area.

Research on models and other analytical tools is important for understanding, predicting, and evaluating the performance of spatial systems. At the same time, research on changes in a spatial system is essential to guide the development of useful tools and to provide a theoretical and logical foundation for the choice and application of tools in a particular spatial setting. A good example is the question of the applicability of monocentric urban models for the analyses of current socio-economic issues in metropolises where the role of CBD has been diminished by the increasing economic role of suburban centers (Richardson 1988)[210]. Thus, it is critically important to understand changes in functions, forms, and structures of spatial systems to develop sound prescriptions for problem solving.

12.2.1 Major Characteristics of Urban and Regional Systems

Let's start with urban systems. What do we understand about the major characteristics that affect the shaping of forms, structures, and functions of modern urban areas that are distinctively different from rural areas? Some of the major urban characteristics may be summarized as follows (Kim 1989)[136]:

1. Land is used intensively in urban areas. The intensive use of land connotes a higher density of population and employment, which in turn causes land and transportation congestion. A corollary to the notion is that congestion is not necessarily an evil to avoid, but a result of a spatial and other scarce resource allocation procedures (Boyce and Kim 1987) [41]. Thus, elimination of congestion should not be a goal of transportation policy; rather, efficient and equitable distribution of resources should be the public sector's goal for an urban area.

2. Both private and public decision makers affect shaping of urban functions, forms, and structures. In urban areas in a mixed economic system, some goods and services are produced by the private sector and some by the public sector as mentioned earlier. The public sector also imposes regulations on almost all private activities in modern cities through zoning, subdivision regulations, building codes, and other land use regulations and pricing.

3. Inputs to urban industries include not only people, but also primary and intermediate goods, making urban economic activities complex and diverse. Transportation systems in urban areas need to provide services not only for the movement of people, but also for moving raw materials, intermediate goods, and finished products.

4. Most urban areas have been built over decades, if not centuries. The current urban form and structure of a given urban area is the result of incremental additions, alterations, and demolitions by both man and nature.

In the context of regional systems, three characteristics stand out.

1. As in urban areas, the interaction between the public and private sector plays a significant role in shaping the socio-economic function and structure of a region. Compared to the situation in urban areas, the public sector usually assumes the stronger role as the leader in regional settings, influencing decisions made by the private sector. This is particularly true in transportation, because the public sector provides new transportation systems, improves and maintains their capacities, and regulates services and prices. A corollary to this notion is that a single objective function such as those we frequently use in conventional programming models may not adequately represent interests of both

public and private sectors. In fact, many times the private sector's objectives may be different, if not conflicting, with those of the public sector.

2. Proper boundaries for socio-economic activities in a region may not be the same as political jurisdictions, because political boundaries were chosen largely for historical and political reasons rather than economic reasons.

3. The economic competitiveness of comparative advantages inherent to a particular region may change over time as technology changes. Today, for example, being located near to mineral resources in most cases brings no comparative advantages, for firms in these regions are no longer economically competitive after the advent of high-tech industries.

12.2.2 Changes in Technology and Lifestyles

Female labor force participation, the declining birthrate, changing household size, the changing economic status of particular social and ethnic groups, and continuing decentralization of economic activity from the traditional CBD are a few examples of changes in lifestyles that will have a profound impact on the future form, function, and structure of urban and regional systems. Undoubtedly, these changes, along with changes in demographics and social patterns, will affect urban locational behavior and travel demand in urban areas.

Changes in technology will also have profound impacts both on lifestyles and changes in functions and structures of urban and regional systems. Several computer companies, including IBM and Kurzweil Applied Intelligence Inc. are developing computers known as "talkwriters" that understand spoken words and then print them on a screen. Having an electronic map on the dashboard of a car is no longer improbable because of technical problem, but is now unusual only because of its cost. Some of emergency vehicles in California have a device on their dashboards that shows the driver how to get to destinations. Currently, the device cost around $3,000. Recent developments in super-conductivity may speed up the development of trains that go 250 mph or more, as fast as some planes. High-speed trains called "MAGLEVS," which levitate on a cushion of magnetism, are being developed in Germany and Japan (*USA Today* December 28, 1988).

Easy access to telecommunication facilities, including microcomputers, facsimile machines, and picture-phones may also affect urban locational and travel behavior. Some travel may be substituted by using one or more of these telecommunication means. People who do not need face-to-face contacts in their production activity may choose a location where access to and from the CBD by a traditional mode of transportation may not be easy. Using data from the Nationwide Personal Transportation Study (NPTS) conducted by the U.S. Department of Transportation for 1977 and 1984, Kumar (1989)[153] showed that commuting distances by workers in information-dominated industries in metropolitan areas have been declining, suggesting that some of the work trips might have been substituted by telecommuting.

In fact, two decades of rapid advances in information and computer technology led some analysts to claim that mankind is now entering to an "information society" (Bell 1979[26]; Benigre 1986[29]; Rogers 1986[215]). With the growing use of telecommunication systems, it is becoming possible for different productive activities such as manufacturing, distribution, management, accounting, and inventory control to be located at different places and yet operated efficiently as a single enterprise (Kumar 1989)[153]. This emerging trend may be represented by new forms of economic agglomerations in a noncontiguous manner. Coopers and Lybrand estimated that electronic messages will soar from $2.4 billion in 1988 to $8 billion in 1991 (*USA Today* December 28, 1988).

Based on these characteristics, which seem to significantly affect the shape of urban and regional functions and structures, this discussion will now assess three emerging research issues in transportation:

1. a supply side view of transportation systems;
2. a demand side view of transportation and spatial interaction; and
3. a technological development view of transportation and telecommunication.

12.3 Transportation Systems: A Supply Side View

There are several excellent works available that provide the basic framework for understanding the supply side of transportation services, including Morlok (1976)[184], Manheim (1980)[170], and Florian and Gaudry (1980)[97]. While there are various views of the supply side of transportation, Turnquist (1985)[249] summarized that the authors' studies agree that the supply side of transportation involves both the technology of providing services and the management actions of the suppliers.

An excellent collection of reports and a summary of the state of the art on research opportunities for transportation systems can be found in a special issue of *Transportation Research* (Boyce 1985)[37]. In this issue, the results of the March 1985 Evanston Workshop on Transportation Research are reported, including assessments on recent developments in transportation systems and discussions of research opportunities in the following three subareas. These are (1) transportation system characteristics and operations, (2) transportation logistic systems, and (3) transportation networks and facilities.

Rather than duplicating what has already been carefully analyzed and suggested, a brief summary of research opportunities identified by the Evanston Workshop participants is introduced and some methods are developed in this chapter that may provide solutions to those research issues.

12.3.1 Transportation System Characteristics and Operations

Two major research issues were suggested by the participants of the Evanston Workshop in 1985 as fundamental research on transportation system characteristics and operations (Boyce 1985, pp.355-356)[37]:

1. As a result of regulatory reform and major advances in computer and communications technology, a major opportunity now exists for the redesign of entire freight transportation systems. Research issues include data sensing and processing, control of complex systems, integrated design of facilities-vehicles-operations-control, and management and organizational structures.

2. As a result of major advances in sensing technology, and computers and communication technology, opportunities exist for the development of advanced vehicle control and guidance systems, and scheduling of vehicle operations. Research issues include (1) dynamic route guidance technology and methods, (2) ergonomic design, (3) driver acceptance, and (4) real-time data collection and control.

As for the issue of the freight transportation system, particularly of urban transportation systems, an additional important point to note is that we have focused on person trip movements for the past thirty years, and have paid little attention to understanding the urban goods movement. An alternative transportation planning model will be developed in a later section that includes both passenger and freight movements in an urban area.

In regard to advanced vehicle control and guidance and scheduling of vehicle operations, only a few authors paid attention to the problems of vehicle fleet sizing and empty vehicle distribution for trucks and rail cars (Jordan and Turnquist 1983[130]; Haghani and Daskin 1984[112]; Jordan and Burns 1984[129]; Powell, Sheffi, and Thiriez 1984[205]). The problem of vehicle size is more acute in developing countries, where overloading practice is sometimes beyond imagination.

Implications to transportation planning of such overloading practice is that the tonnage shipped and the maximum capacity of vehicles passing through certain checkpoints diverges widely. At the same time, network capacity investment planning based on the number of vehicles forecast (not considering vehicle fleet size which accounts for the overloading) would result in smaller amounts of capacity increase. In addition, overloading not only harms vehicles, but also damages the road systems. In fact, improvement of road conditions has become a major concern for many developing countries in the 1980s. In many developing countries, roads have deteriorated in the past decade due to their carrying heavier traffic loads than were originally intended for them, and funds have been allocated for expanding rather than maintaining the network (World Bank 1983)[13].

A research team at the Korean Institute of Science and Technology (KIST) has recognized the important problems of dealing with vehicle fleet size and empty car backhaul, and has developed models for vehicle fleet size, truck backhaul, and empty rail car distribution (Kim 1989)[136]. The models have been created as part of the national transportation planning model for the development of the Fifth Five-Year Development Planning programs in Korea (Kim and Kim 1985).[140] These models are reintroduced in the Appendices.

12.3.2 Transportation Logistic Systems

Transportation logistics can be defined as the study of how manufactured goods overcome time and distance in the process of being transformed from raw materials to consumed products (Daskin 1985)[77]. Logistics are an important component of the economy, as can be seen in the U.S., where logistics costs have been estimated at between fifteen and twenty-three percent of the GNP, depending on whether or not service industries are included in the GNP.

The participants in the Evanston Workshop in 1985 identified general issues in transportation logistics as: (1) integration of transportation and production, (2) evaluation of classes of methods and their application context, and (3) the relationship of logistics with the outside world.

In many logistical studies, the emphasis has been on the interaction of shippers and carriers and their decision- making processes. The possibility of modeling the interaction of shippers and carriers in bilevel programming frameworks has been suggested and new findings in algorithms for solving a large-scale bilevel programming problems are reported in Suh and Kim (1989)[238]. A general description of a bilevel programming problem is introduced in the later section of this chapter.

Expert systems, a branch of artificial intelligence, has emerge as a promising tool for developing integrated models of several logistical functions such as location, fleet selection, inventory, production, marketing and vehicle routing. An interesting application of an expert system is to search for the optimal routing interactively and solve routing problems in real time by combining an expert system with an optimization routing model (Golden and Baker 1985)[109]. A detailed explanation of the possibility of combining expert systems with other existing tools will be introduced in a later section.

12.3.3 Transportation Networks and Facilities

The Evanston Workshop in 1985 concluded with the following recommendations regarding research opportunities on transportation networks: (1) formulation and solution of multilevel decision models; (2) stochastic, dynamic, and disequilibrium extensions of models; (3) the study of existence , uniqueness, sensitivity, and stability of solutions; (4) the error analysis of models; (5) statistical estimation of spatial interaction and demand model parameters; (6) development of more efficient methods for large models; and (7) network aggregation and decomposition.

This chapter will not summarize the work on transportation networks, as excellent reviews have been published elsewhere (Fernandez and Friez 1983[90]; Boyce 1984[38]; Friez 1985[100]). Instead, in a later section a bilevel network design model is proposed that may shed light on some of the research issues identified above and that explicitly explains the relationship between travel and location choices and between economic productivity and the design of transportation networks. Such a model was also developed in a interregional setting and applied to an evaluation of highway investment strategies for Korea by Suh and Kim (1989)[237].

Research opportunities on transportation facilities can be highlighted by citing the following statistics in the U.S. About forty-four percent of 565,000 bridges on the U.S. street and highway system are considered deficient. There are about 3.9 million miles of streets and highways and about 266,000 miles of railroad lines in which deficiencies have been developing at a rate faster than funds have become available to maintain it. Current estimates of costs to repair, rebuild, and upgrade public works in which transportation facilities of all kinds account for a major portion of these needs range from one trillion to three trillion dollars over the next twenty years (Moavenzadeh 1985)[182]. Research issues and problem areas in transportation facilities are well analyzed in both papers by Moavenzadeh (1985)[182] and Garrison (1985)[105].

12.4 Transportation and Spatial Interaction: A Demand Side View

Among urban and regional systems models developed thus far, only a few models have explicitly addressed some of the important characteristics of urban areas identified above. Mills (1972) [178] first introduced a model that explicitly analyzes the intensive use of urban land employing a three-dimensional input-output model. He also introduced a method to analyze transportation congestion first conceptually (Mills 1972) [178] and then numerically (Mills 1976) [179] using the integer programming technique. Transportation networks in Mills' models, however, are abstract forms of real-world transportation networks. The public-private interaction in Mills' models is represented in an optimization framework that permits calculation of an optimum allocation of both public and private resources. Further, Mills' models demonstrated that competitive markets can sustain an optimum allocation of resources if the public sector provides its services in optimum fashion.

A method for determining the transportation service requirement for both passengers and freight combined was first introduced by Leontief and Strout (1963) [164] in a regional context and later expanded by Wilson (1970a[260], 1970b[259]). In these models, Wilson used an input-output structure for the production sector and entropy maximization for the transportation sector.

Erlander (1977) [85] introduced the notion of "dispersion factors," which can now simulate the fact that a city has been incrementally built and that the present form and structure of a city, have deviated from its optimum form and structure. He used entropy as a constraint instead of maximizing it, compared to Wilson (1970a) [260] who used it in objective functions. Kim (1979 [134], 1983[135], 1986[137], 1989[136]) introduced an extended model with a nonlinear structure by combining Wilson's commodity flow model, Erlander's dispersion factors, and Mills' urban activity model.

While past and current research on combined models of urban activities has been limited, there has been a considerable amount of related research that provides a basis for building combined urban activity models and developing efficient solution algorithms. Figure 12.1 includes a brief description of previous studies that highlights some of the important developments in building an integrated urban systems model.

Of the existing urban systems models that are formulated to guide public sector planning for future urban growth, most are inadequately articulated to permit determination of whether market solutions are efficient and, if not, what kind of public intervention would be best to ensure efficiency. Such models derive the urban spatial patterns from conditions of supply and demand, profit and utility maximization, and market equilibrium without optimality criteria.

Improvements have been made in models by Mills (1972[178], 1976[179]) and Kim (1989) [136] in handling public-private interaction. Even in the these models, however, a single objective criterion was assumed to represent society's welfare despite the fact that the objectives of the public and private sectors are frequently different, if not conflicting. While the need for the valuative criteria is prevailing for the assessment of public policies, the applications of a single valuative criterion is increasingly complex. After all, one of the critical questions in a mixed economic system is how much the public sector should intervene in order to enhance the market efficiency. Simultaneous representation of different and/or conflicting objectives in building urban and regional systems models should be the subject of continued research.

Although the rational decision-making models have provided useful recommendations for public policy, it is important to recognize that there are other useful perspectives on decision making, particularly on public decision-making behavior. Linstone (1984)[166] argues that alternative perspectives to the rational decision-making procedure exist, that viewing problems through several perspectives is like viewing an object through several different lenses, and that multiple objectives can clarify the analysis of any area of policy planning. Allison (1971) [5] introduced the notion of multiple per-

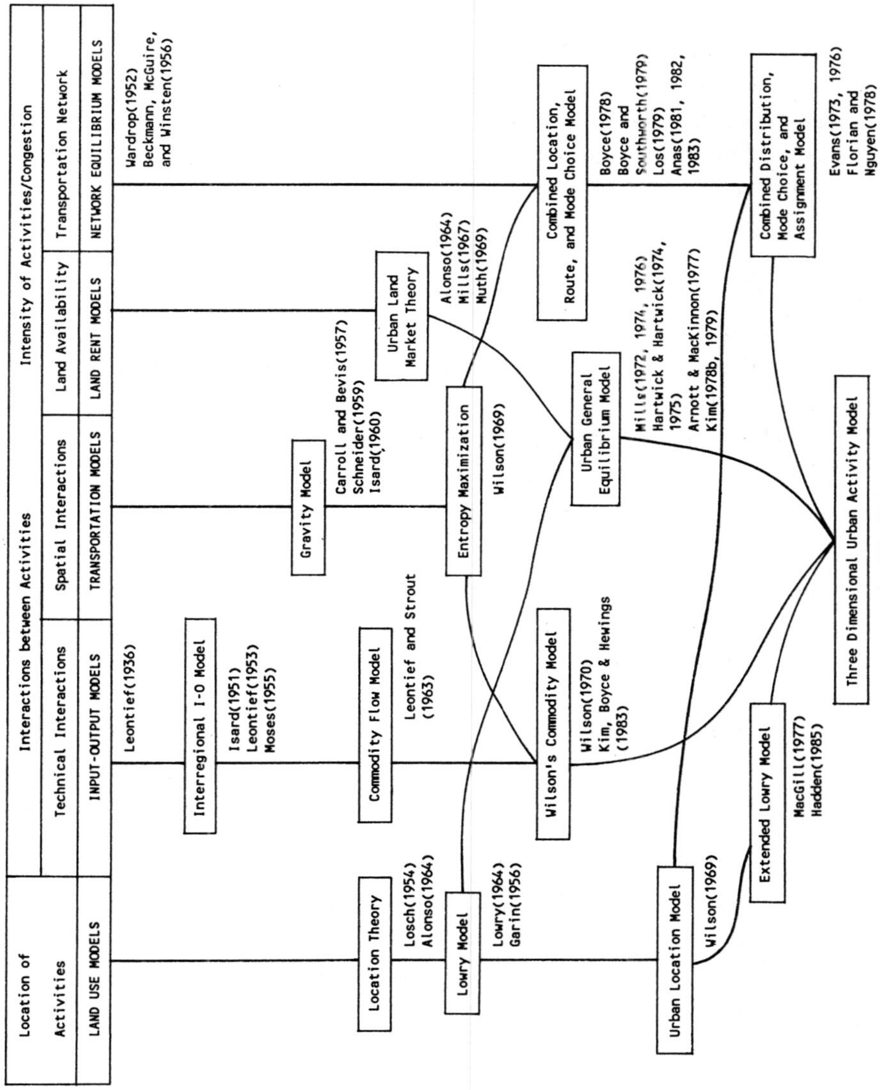

Figure 12.1: Developments in Urban and Regional Systems Models

spectives in policy making; and Steinbrunner (1974) [233] applied the same basic concept to an analysis of planning a proposed multinational force for NATO.

Wachs (1985) [250] recognized that one of the major attributes of organizational perspectives of decision making is that there are multiple decision making units, each pursuing its own goals, and each interested in enhancing its status. As described previously, there are at least two actors making decisions affecting urban systems in a mixed economic system: the public and private sectors. Within public sectors, there are many different units making decisions that affect urban systems: the federal government, the state governments, and various local units.

The foregoing analyses strongly suggest the formulation of urban systems models in a multilevel programming framework. A special case where only two different decision units are involved is called a bilevel programming problem, the nature and structure of which have been introduced in the previous chapter.

While development of efficient solution algorithms should be the subject of continued research, the multilevel programming approach has many features that other techniques do not have.

1. Different, if not conflicting, objectives of two or more decision makers and/or different units can be simultaneously analyzed. It implies that the public-private interaction explicitly can be evaluated.

2. Multiple valuative criteria can be explicitly included in a model, compared to a single valuative criterion in conventional programming approaches. In this view, a multicriteria decision-making method can provide realistic policy recommendations.

3. The multilevel programming approach is a conflict resolution method because the approach can accommodate conflicting objectives. Many conflicting public objectives, such as economic development and environmental preservation in an urban area, can now be evaluated in terms of broader societal goals.

Further research on bilevel and multilevel programming approaches and their applications to urban and regional systems analysis is one area in which research promises major payoffs.

12.5 Transportation and Telecommunication

The salient characteristic of the computer, its enormous capacity and speed in storing, accessing, and processing data, has made it an indispensable tool for urban decision makers who deal with varied information for their problem-solving tasks.

Recent advances in artificial intelligence (AI) technology raise interest in how existing urban systems modeling techniques might benefit from AI. Artificial intelligence is not a technology that solves urban problems by itself. Rather, it is a newly emerging and promising technology that can be incorporated or integrated into the existing urban systems models to provide more intelligent and effective solutions to urban problems (Kim, Wiggins, and Wright 1989)[143]. Ongoing AI research shows the possibilities of developing an intelligent computer that is able to reason, learn, and understand human language. As decision makers' tasks become more complex, the role of computers and computerized information systems becomes increasingly important to urban systems analysis (Han and Kim 1989) [113].

In the face of uncertainty and dynamic changes in urban society, the rational decision-making model alone may not be adequate for responding to continuously challenging urban problems. Indeed, the model may have to be supplemented by additional perspectives based on the following attributes of decision makers' behavior in an organization (Wachs 1985)[250]:

1. Short-term consequences of decisions are emphasized over long-term consequences.

2. Innovation is less important than organizational security and enhancement.

3. Preferred policies are those most easily absorbed into an organization's on going values and programs.

4. There is greater reliance on relationships than on formalized technical information.

5. Decisions are preferred on the basis of the extent to which they enhance one's role in an organization or one's standing in a profession.

6. Intuition and rules-of-thumb are at least as important as analysis.

7. Memory and personal experience are more important than technical analysis.

8. Decision processes focus on simple hypotheses and a few important alternatives.

Few would dispute the importance of these decision-making attributes in developing alternative strategies for urban problem solving. While the conventional spatial systems modeling techniques cannot handle these important attributes, expert systems, a branch of AI, may be able to do so. While it is beyond the scope of this chapter to describe expert systems in detail, existing urban systems models can be attached to a new expert system in which rule-based reasoning and qualitative simulation are tied into quantitative systems analysis.

In general, the expert system is defined as "a computer system that uses a representation of human expertise in a specialty domain in order to perform functions similar to those normally performed by a human expert in that domain" (Goodall 1985) [110]. The essence of expert systems is that they attempt to incorporate the judgment, experience, rules-of-thumb, and intuition of human experts into problem solving.

The efforts made by several authors to distinguish expert systems from conventional programming are useful in understanding the nature of expert systems. According to Waterman (1986) [253], "an algorithmic method of conventional programming is designed to produce optimal solutions, whereas the heuristic method of expert systems produces an acceptable solution most of the time." Thus, the problems suitable for expert system development must require heuristic solutions rather than algorithmic solutions. Such problems in urban systems analysis may include finding acceptable alternatives while considering those attributes described above or settling a lawsuit in which heuristics and rules-of-thumb can effectively limit the search for solutions.

As depicted in Figure 12.2, combining expert systems and urban systems models requires (1) a user interface, which allows the user to communicate with the system, providing necessary data to the system; (2) an inference engine, which solves given problems using input data from the user and knowledge from the knowledge base, through its own reasoning methods; (3) a knowledge base and qualitative analysis that contain the knowledge obtained from a domain expert, including facts, beliefs, rules-of-thumb, and other judgmental factors of the human expert; and (4) urban systems models.

The inference engine, the top layer of the structure, uses a rule-based paradigm to process domain knowledge that is predominantly declarative. The mid-layer uses qualitative causal structure descriptions to express the

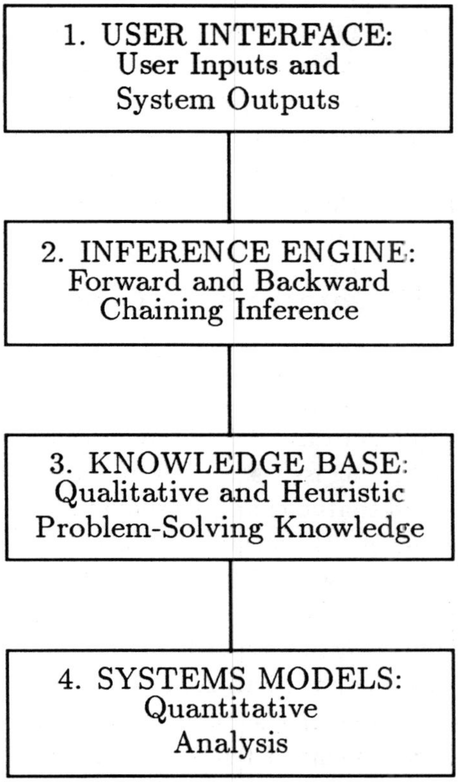

Figure 12.2: Linking Urban Systems Models with Expert Systems

relationships among decision parameters. The bottom layer is made up of the conventional systems models that are quantitative in nature.

Decision makers are often dissatisfied with the usefulness of systems approaches not so much because the work is inconsistent or technically limited, but because it appears misguided and irrelevant. Decision makers and public organizations do not necessarily behave in accordance with a rational model; because they don't, one important future research item for urban and regional systems modeling is to combine true decision makers' behavior with a rational systems model. Developing and combining expert systems and urban systems models should be the subject of continued research.

12.6 Toward Developing a Research Agenda

This chapter has attempted to assess current developments in transportation, spatial interaction, and telecommunication in order to shed light on emerging research opportunities. Discussions have been centered around identifying old and continuing issues, and particularly those issues connected to the transportation and spatial changes brought about by the new telecommunication technologies. For the former issues, a bilevel programming model has been introduced that explicitly includes (1) the public and private interaction, (2) three-dimensional land use and socio-economic urban activities, and (3) transportation congestion caused by the simultaneous consideration of moving both passengers and urban goods.

As for the latter category, major impacts on transportation and spatial interaction by the advancement of telecommunication technologies may be narrowed to two: impacts on spatial flexibility of socio-economic activities; and increases in efficiency in planning, management, operation, maintenance, and scheduling of transportation facilities and systems.

The former impact poses further challenging research questions: Would the advances in telecommunication technologies allow significant opportunities for increased flexibility in the locating economic activities, as was alluded to by several authors, including Castells (1985)[58], Nicol (1985)[196], Stanback (1985)[232], and Brotchie (1984)[50] ? Would communication modes allow separation of workers from work materials, substituting for physical travel, as analyzed by Goddard (1980)[108] and Kumar (1989)[153] ?

Regarding the latter question, major advances in artificial intelligence, particularly in expert systems have created many challenging research opportunities for the future (Han and Kim 1989)[113]. In addition to the concept of combining expert systems with operations research models, there are other potential combinations that merit regional scientists' further attention.

1. Combining expert systems (ES) and database management systems (DBMS) is one area of research whose time has come. The idea of combining ES and DBMS has received a great deal of attention recently, creating another hybrid information system called the "expert database system" (EDBS). The marriage of ES and DBMS takes two fundamental forms. The first type of coupling is with "an expert" or "intelligent" interface to a standard DBMS or an "expert" query optimizer with a standard DBMS. The second type of coupling includes a standard expert system integrated with a large relational database of facts (Brachman 1987)[45].

The functions of the intelligent interface may include formulating efficient queries for the user by incorporating the knowledge of a human expert in the domain of the database, or interpreting data retrieved from the database and eliminating any inconsistencies (for instance, unifying the different units of measurement used in the database). The functions of the database system in the second type of coupling include the direct use of variables in the database system by the rules in the expert system. In fact, the inability of an ES to access or utilize a database already developed has been one of the major deficiencies of ES.

Goodall (1985) points out some possible difficulties with the coupling of ES and DBMS: (1) an ES and a database system are both big programs, making it difficult to run both together; (2) database systems may not be able to answer all the forms of query that an ES would put to them, and (3) an ES may not understand the replies of a database system.

But these problems have been solved recently by the introduction of several ES shells, including GURU from Micro Database Systems, Inc., which provides a database manager fully integrated into an ES (Micro Database Systems, Inc. 1987)[177].

2. Combining ES and geographic information systems (GIS) is another promising area of research. The major function of GIS is to store and manipulate spatial data for cartographic modeling. The basic need for coupling ES with GIS stems from the fact that many tasks involved in cartographic modeling require expertise on the part of the user in the particular subject matter, and that expertise can be provided by experts or expert systems.

For example, the task of identifying suitable sites for a particular land use can be effectively assisted by GIS, which easily identifies the area with desired or undesired characteristics through several overlay manipulations. Another use of expert systems in GIS may be to provide a useful and more intelligent interface to the user. ES may be designed to help users devise efficient operating procedures for cartographic modeling and interpret the results of spatial analysis. ES can also be used in GIS for uncertainty reasoning and to resolve inconsistent and contradictory information obtained from GIS. Among the problems of integrating ES with GIS, Robinson, Frank, and Blaze (1986)[214] point out the limited capability of current ES shells and the lack of formalism typical in geography.

3. A third field involves combining expert systems (ES) and decision support systems (DSS). The idea of integrating ES into DSS to create more powerful and useful computer-based systems has been given much attention recently, creating a new terminology, "expert decision support system" (EDSS), or "intelligent decision support system." The possible contributions of ES in EDSS include (1) helping users in selecting models, (2) providing judgmental elements in models, (3) simplifying building simulation models, (4) enabling friendlier interface, and (5) providing explanation capability (Turban and Watkins 1986).

Most of all, the ES can play an important role in EDSS with model selection and building. As Strauch (1974)[235] points out, the process of problem analysis (or modeling) usually involves three interrelated components: formulation of the formal problem, mathematical analysis, and interpretation of the results. While the mathematical analysis is handled well with DSS, the formulation requires the subjective knowledge of the user. Further, the interpretation requires the personal judgment of decision makers. The coupling of ES and DSS in this case is based on the assumption that subjective knowledge and personal judgment can be best made by experts.

An example of the use of ES for this purpose is the Advisory System for Ground Water Quality Assessment (Armstrong 1987)[8], which maintains a knowledge base in addition to the model base to assist decision makers in determining proper sample patterns and size given a specified confidence level and evaluating samples obtained. Wood and Wright (1987)[264] also show an example of adding a rule-based system to the simulation model for storm water management for the purpose of aiding users with model calibration and result interpretation.

The intelligent interface provided by ES for the modeling tasks in DSS may stimulate planners to employ mathematical models more frequently and easily in their problem-solving processes. As often criticized by planners, modeling components tend to be treated as a black box; they are inadequate in recognizing the need for judgments by the users. Further, they often conceal implicit judgments and assumptions by the users (Langendorf

1985)[154].

The study by Wellman (1986)[254] that attempts to connect ES to mathematical modeling is worth much attention. He reports a rule-based system that generates parameters for the user to make the mathematical models easy to use. In that system, ES serves as an extra layer between the model and the user, translating qualitative criteria into the numeric input and also translating the model's numeric output to qualitative concepts that are more intuitive and informative to the user (Wellman 1986)[254]. This approach certainly is a great improvement over the unaided use of modeling algorithms, encouraging easier use of quantitative modeling to support many planning decisions.

In summary, the coupling of ES and DSS basically takes two different forms: integration of ES into the conventional DSS to provide qualitative reasoning capability and intelligent user interface; and integration of DSS into the conventional ES to provide modeling capability. In the first type of coupling, ES may help the users select proper models, input necessary parameters, and interpret outputs of DSS. In the second type of coupling, DSS provides modeling capability to ES, recognizing that human experts often use quantitative models to support their experience, intuition, or rules-of-thumb. For example, site selection tasks require planners' intuition and experience to develop selection constraints and criteria, but the task of selecting the best site may require planners to use multiobjective optimization modeling to make more competent decisions (Suh, Kim and Kim 1988[239]; Han and Kim 1989[113]).

Appendix A

Optimality Conditions

The Lagrangian for the model shown by Equations (6.2) through (6.8) is

$$\begin{aligned} L &= \sum_a \int_0^{f_a} c_a(x)dx + \sum_i \sum_r p_r^i X_r^i + \sum_{i\varepsilon e} \sum_r d_r^i (E_r^i + I_r^i) \\ &+ \sum_r \sigma_r (E_r - \sum_i E_r^i) + \phi_r (I_r - \sum_i I_r^i) \\ &+ \sum_i \sum_r \alpha_r^i (\sum_q a_{rq} X_q^i + \sum_j \sum_p X_r^{ijp} + E_r^i - \sum_j \sum_p X_r^{ijp} - X_r^i - I_r^i) \\ &+ \sum_r \frac{1}{\beta_r} [S_r + \sum_i \sum_j (\sum_p X_r^{ijp}) ln (\sum_p X_r^{ijp})]. \end{aligned}$$

To determine the optimality conditions when $X_r^{ijp} > 0, X_r^i > 0, E_r^i > 0$ and $I_r^i > 0$, differentiate L with respect to each of these variables:

$$\frac{\partial L}{\partial X_r^{ijp}} = \sum_a c_a(f_a) \sigma_a^{ijp} g_r + \alpha_r^i - \alpha_r^j + \frac{1}{\beta_r} [1 + ln(\sum_p X_r^{ijp})] = 0 \quad \text{(A.1)}$$

$$\frac{\partial L}{\partial X_r^i} = p_r^i + \sum_q \alpha_q^i (a_{rq}) - \alpha_r^i = 0 \quad \text{(A.2)}$$

$$\frac{\partial L}{\partial E_r^i} = d_r^i - \sigma_r + \alpha_r^i = 0 \quad \text{(A.3)}$$

$$\frac{\partial L}{\partial I_r^i} = d_r^i - \phi_r - \alpha_r^i = 0. \quad \text{(A.4)}$$

To determine optimality conditions for non export/import zones i, first define $\sum_a c_a(f_a) \delta_a^{ijp} g_r \equiv c_r^{ijp}$. Since the shipment cost of all routes chosen between i and j for the shipment of r is the same at the equilibrium, $c_r^{ijp} = c_r^{ij}$ at the optimum solution. Thus, for all non-export/import zones i, Equations (A.1) and (A.2) can be expressed as

$$\sum_p X_r^{ijp} = \exp \quad [\beta_r(\alpha_r^j - \sum_q \alpha_q^i(a_{rq}) - p_r^i - c_r^{ij})] \tag{A.5}$$

which is equation (6.11).

Appendix B

Standardization of Units from the 1978 Survey

The units in the 1978 origin-destination survey were tons/day for freight and persons/day for passengers. The following are assumptions in converting these results to standard units:

1. Five percent of freight and fifteen percent of passenger flows are assumed to be one-hour flows.

2. Tons were converted to number of trucks assuming 1.5 tons/truck, which in turn were converted to passenger car units (PCU) by three PCUs/truck.

3. It was assumed that one bus carries thirty passengers and one bus is assumed to be equivalent to three PCUs.

4. Car occupancy ratio was assumed to be 2.7 persons/car.

5. Link capacity was estimated based on the following:

$$S_a = P_a \cdot NL$$

S_a: the capacity of link a

P_a: uncongested capacity per lane by facility type:

Unpaved highway 1,720 PCUs/hour/lane
Paved highway 2,125 PCUs/hour/lane
Expressway 2,500 PCUs/hour/lane

NL: number of lanes

Bibliography

[1] H.Z. Aashitiani and T.L. Magnanti. Equilibria on a congested transportation network. *SIAM Journal on Algebraic and Discrete Methods*, 2:213–226, 1981.

[2] M. Abdulaal and L. LeBlanc. Continuous equilibrium network design models. *Transportation Research*, 13B:19–32, 1979.

[3] Automic Energy Research Establishment (AERE). *Harwell, Subroutine Library Specification: VF03, VA13 and VE03.* Harwell, England, December 1985.

[4] Y. Ahmed et al. *Road Investment Programming for Developing Countries: An Introduction Example.* The Transportation Center for Northwestern University, Evanston, IL, 1976.

[5] G. Allison. *Essence of Decision: Explaining the Cuban Missile Crisis.* Little, Brown, Boston, 1971.

[6] R.E. Allsop. Some possibilities for using traffic control to influence trip distribution and route choice. In D.J. Buckley, editor, *Transportation and Traffic Theory*, pages 345–375, Sixth International Symposium on Transportation and Traffic Theory, Elsevier, New York, 1974.

[7] L. Armijo. Minimization of functions having lipschitz continuous first partial derivatives. *Pacific Journal of Mathematics*, 1:1–3, 1966.

[8] M.P. Armstrong. A rule-based advisory system for ground water quality assessment at hazardous waste disposal sites. 1987. Department of Geography and Computer Science, The University of Iowa.

[9] A.A. Assad. Multicommodity network flows: a survey. *Networks*, 8(1):37–92, 1978.

[10] M. Avreil. *Nonlinear Programming: Analysis and Methods.* Prentice-Hall, Englewood Cliffs, NJ, 1976. Series in Automatic Computation.

[11] M. Bacharach. *Bi-proportional matrices and input-output change.* Cambridge University Press, New York, NY, 1970.

[12] E. Balas. An infeasibility pricing decomposition model for linear programs. *Operations Research*, 14:847–873, September-October 1966.

[13] World Bank. *World Development Report.* Oxford University Press, Oxford, United Kingdom, 1983.

[14] G.M. Barber. A mathematical programming approach to a network development problem. *Economic Geography*, 51(2):128–141, 1975.

[15] G.M. Barber. Sequencing highway network improvements: A case of south Sulavei. *Economic Geography*, 53(1):55–69, 1977.

[16] J.F. Bard. An algorithm for solving the general bi-level programming problem. *Mathematics of Operations Research*, 8:260–272, 1983.

[17] J.F. Bard. Coordination of a multidivisional organization through two levels of management. *OMEGA*, 11(5):457–468, 1983.

[18] J.F. Bard. An efficient point algorithm for a linear two-stage optimization problem. *Operation Research*, 31:670–684, 1983.

[19] J.F. Bard. Geometric and algorithmic development for a hierarchical planning problem. *European Journal of Operation Research*, 19:372–383, 1985.

[20] J.F. Bard. A grid search algorithm for the bilevel programming problem. *Proceedings of 14th Annual Meeting of the American Institute for Decision Science*, 256–258, 1982.

[21] J.F. Bard. Optimality conditions for the bilevel programming problem. *Naval Research Logistics Quarterly*, 31:13–26, 1984.

[22] J.F. Bard and J.T. Moore. A branch and bound algorithm for the bilevel programming problem. 1987.

[23] M.J. Beckmann. Principles of optimum location for transportation networks. In W. Garrison and D. Marble, editors, *Quantitative Geography*, page Part I: Economic and Cultural Topics, Northwestern University, Evanston, IL, 1967.

[24] M.J. Beckmann, C.B. McGuire, and C.B. Winsten. *Studies in the Economics of Tranportation.* Yale University Press, New Heaven, CT, 1956.

[25] C. Bell, P. Hazell, and R. Slade. *Project Evaluation in Regional Perspective.* Johns Hopkins University Press, Baltimore, MD, 1982.

[26] D. Bell. Communications technology for better or for worse. *Harvard Business Review*, 20–42, 1979. May-June.

[27] O. Ben-Ayed. *Bilevel Linear Programming: Analysis and Application to the Network Design Problem.* PhD thesis, University of Illinois at Urbana-Champaign, Urbana, IL, 1988.

[28] O. Ben-Ayed, D.E. Boyce, and C.E. Blair. A general bilevel linear programming formulation of the network design problem. *Transportation Research*, 22B(4):311–318, 1988.

[29] J.R. Beniger. *The control revolution.* Harvard University Press, Cambridge, MA, 1986.

[30] G. Bergendahl. A combined linear and dynamic programming model for interdependent road investment planning. *Transportation Research*, 3:211–228, 1969.

[31] D.P. Bertsekas. *Constrained Optimization and Lagrange Multiplier Methods.* Academic Press, New York, NY, 1982.

[32] W.F. Bialas and M.N. Chew. A game-theoretic approach to coalition formation in multi-level decision-making organization. 1981. Operations Research Program.

[33] W.F. Bialas and M.H. Karwan. Multi-level linear programming. 1978. Operation Research Program.

[34] W.F. Bialas and M.H. Karwan. Two level programming. *Management Science*, 30(8):1004–1020, August 1984.

[35] J. Bisschoop, W. Candler, J.H. Duloy, and G.T. O'mara. The indus basin model: a special application of two-level linear programming. *Mathematical Programming Study*, 20:30–38, 1982.

[36] D.E. Boyce. Integration of supply and demand models in transportation and location: Problem formulation and research questions. *Environment and Planning A*, 18:485–489, 1986.

[37] D.E. Boyce. Transportation research: the state of the art and research opportunities. *Transportation Research*, 19A, 1985.

[38] D.E. Boyce. Urban transportation network-equilibrium and design models: Recent achievements and future prospects. *Environment and Planning A*, 16:1445–1474, 1984.

[39] D.E Boyce and G.J.D. Hewings. Interregional commodity flow, input-output and transportation modelling: and entropy formulation. 1980.

[40] D.E. Boyce, B.N. Janson, and R.W. Eash. The effect on equilibrium trip assignment of different link congestion functions. *Transportation Research*, 15A:223–231, 1981.

[41] D.E. Boyce and T.J. Kim. The role of congestion of transportation networks in urban location and travel choices. *Transportation*, 14:53–62, 1987.

[42] D.E. Boyce, L.J. LeBlanc, and K.S. Chon. Network equilibrium models of urban location and travel choices; A retrospective survey. *Journal of Regional Science*, 28(2):159–183, 1988.

[43] D.E. Boyce and J.L. Soberanes. Solutions to the optimal network design problem with shipments related to transportation cost. *Transportation Research*, 13B:65–80, 1979.

[44] D. Braess. Uber ein paradoxon der verkehrsplanung. *Unternehmensforschung*, 12:258–268, 1968.

[45] R.K. Brail. *Microcomputers in Urban Planning and Management.* Center for Urban Policy Research, NJ, 1987.

[46] D. Branston. Link capacity functions: a review. *Transportation Research*, 10:223–236, 1976.

[47] T.A. Broadbent. Activity analysis of spatial allocation models. *Environment and Planning*, 5:673–691, 1973.

[48] M.S. Bronzini. Evolution of a multimodal freight transportation network model. *Proceedings of Transportation Research Forum*, 21:475–485, 1982.

[49] M.S. Bronzini, editor. *Multimodal network model for coal analysis*, Argonne National Laboratory, Argonne, IL, 1980. Proceedings of Coal Transportation Modeling Workshop.

[50] J.F. Brotchie. Technological change and urban form. *Environment and Planning A*, 16:583–596, 1984.

[51] Bureau of Public Roads. *Highway Capacity Manual: Practical Applications of Research.* U.S. Department of Commerce, Washington, D.C., 1950.

[52] Bureau of Public Roads. *Traffic Assignment Manual.* U.S. Department of Commerce, Urban Planning Division, Washington, D.C., 1964.

[53] CACI, Inc. *Transportation Flow Analysis.* Technical Report DOS-OST-P-10-29 to 32, U.S. Department of Transportation, 1980.

[54] E.W. Campbell, L.E. Keefer, and R.W. Adams. A method for predicting speeds through signalized street sections. *Highway Research Board Bulletin*, 230:112–115, 1959.

[55] W. Candler, J. Fortuny-Amat, and B. McCarl. The potential role of multi-level programming in agricultural economics. *American Journal of Agricultural Economics*, 63(3):521–533, 1981.

[56] W. Candler and R. Townsley. A linear two-level programming problem. *Computer and Operation Research*, 9(1):59–76, 1982.

[57] R.G. Cassidy, M.J.L. Kirby, and W.W. Raikes. Efficient distribution of resources through three levels of government. *Management Science*, 17(8):462–473, 1971.

[58] M. Castells, editor. *High Technology, Space, and Society.* Volume 28 of *Urban Affairs Annual Reviews*, Sage Publications, Inc., 1985.

[59] D.D. Cha and W.Y. Kim. *Highway Capacity and its Application to Transportation Planning.* Research Report 86-7, Korea Research Institute for Human Settlements, Seoul, Korea, 1986.

[60] R.M. Chamberlain, C. Lemarechal, H.C. Pedersen, and M.J.D. Powell. The watchdog technique for forcing convergence in algorithms for constrained optimization. *Mathematical Programming Study*, 16:1–17, 1982.

[61] S.C. Choe and T.J. Kim. Spatial distribution of activities in Korea: issues and a research agenda. *Korean Journal of Regional Science*, 1:25–36, 1985.

[62] K.J. Choi. *A Study on Comparative Analysis of Capacity Restraint Function and Method of Application.* Master's thesis, Seoul National University, Seoul, Korea, 1984.

[63] K.S. Chon, Y.J. Lee, K.T. Lin, D.E Boyce, and L.J. LeBlanc. Implementation and computational issues for combined models of location, destination, mode and route choice. *Environment and Planning A*, 15:1219–1230, 1983.

[64] D.H. Chun and K.S. Lee. Changing location patterns of population and employment in the seoul region. 1985.

[65] J.D. Coelho and H.C.W.L. Williams. On the design of land use plans through locational surplus maximization. *Papers of the Regional Science Association*, 40:71–85, 1978.

[66] J.L. Cohon. *Multiobjective Programming and Planning.* Academic Press, New York, 1978.

[67] W.W. Cooper, H.J. Leavitt, and M.W. Shelly. *New Perspectives in Organization Research.* John Wiley and Sons, New York, 1964.

[68] R. Corbis and D. Vallet. An interindustry interregional table of the French economy. In K.R. Polenske and J.V. Skolka, editors, *Advances in Input-Output Analysis*, pages 231 – 249, Ballinger, New York, NY, 1976.

[69] S. C. Dafermos. *Traffic Assignment and Resource Allocation in Transportation Networks.* PhD thesis, The John Hopkins University, Baltimore, MD, 1968.

[70] S. C. Dafermos and A. Nagurney. On some traffic equilibrium theory paradoxes. *Transportation Research*, 18B:101–110, 1984.

[71] S.C. Dafermos. Sensitivity analysis in variational inequalities. *Mathematics of Operations Research*, 13(3):421–434, August 1988.

[72] S.C. Dafermos. Traffic equilibrium and variational inequalities. *Transportation Science*, 14:42–54, 1980.

[73] C.F. Daganzo. On the traffic assignment problem with flow dependent costs I,II. *Transportation Research*, 11:433–441, 1977.

[74] C.F. Daganzo. Some statistical problems in connection with traffic assignment. *Transportation Research*, 11:385–389, 1977.

[75] C.F. Daganzo and Y. Sheffi. On stochastic models of traffic assignment. *Transportation Science*, 11(3):253–274, 1977.

[76] G.B. Dantzig and P. Wolfe. Decomposition principles for linear programs. *Operations Research*, 8:101–111, 1960.

[77] M.S. Daskin. Logistics: An overview of the state of the art and perspectives on future research. *Transportation Research*, 19A:383–398, 1985.

[78] K.B. Davidson. A flow travel time relationship for use in transportation planning. *Proceedings of Australian Road Research Board, Bulletin*, 3(part 1):183–194, 1966.

[79] A.H. de Silva. *Sensitivity Formulas for Nonlinear Factorable Programming and Their Application to the Solution of an Implicitly Defined Optimization Models of US Crude Oil Production.* PhD thesis, George Washington University, Washington, D.C., 1978.

[80] L. Devarajan. A note on network equlilbrium and noncooperative games. *Transportation Research*, 15B:421–426, 1981.

[81] Y. M. I. Dirickx and L. P. Jennergren. *Systems Analysis by Multilevel Methods: With Application to Economics and Management.* John Wiley and Sons, New York, 1979.

[82] J.J. Dongarra, J.R. Bunch, C.B. Moler, and G.W. Stewart. *LINPACK User's Guide.* SIAM, Philadelphia, 1979.

[83] T.P. Egan. *The modeling of coal and transport industry interactions.* REPORT ANL/EES-TM-190, Argonne National Laboratory,

Argonne, IL, 1982. Proceedings of Coal Transportation Modeling Workshop.

[84] S.C. Eisenstat, M.C. Gursky, M.H. Schultz, and A.H. Sherman. *Yale Sparse Matrix Package, I and II.* Research Report 112, 114, Yale University, May 1977.

[85] S. Erlander. Accessibility, entropy and the distribution and assignment of traffic. *Transportation Research*, 11:149–153, 1977.

[86] N.H. Nie et al. *SPSS: A Statistical Package for the Social Science.* McGraw-Hill, New York, third edition, 1975.

[87] S. Evans. Derivation and analysis of some models for combining trip distribution and assignment. *Transportation Research*, 10:35–57, 1976.

[88] S. Evans. *Some applications of mathematical optimization theory in transport planning.* PhD thesis, University College, London, England, 1973.

[89] S. Evans and J. Baxter. Regionalizing national projections with a multiregional input-output linked to a demographic model. *Annals of Regional Science*, 14:52–73, 1980.

[90] J.E. Fernandez and T.L. Friesz. Equilibrium predictions in transportation markets: The state of the art. *Transportation Research*, 17B:155–172, 1983.

[91] A.V. Fiacco. *Introduction to Sensitivity and Stability Analysis in Nonlinear Programming.* Academic Press, New York, NY, 1983.

[92] A.V. Fiacco. Sensitivity analysis for nonlinear programming using penalty methods. *Mathematical Programming*, 10:287–311, 1976.

[93] A.V. Fiacco and G.P. McCormick. *Nonlinear Programming: Sequential Unconstrained Minimization Techniques.* Wiley, New York, 1968.

[94] C. S. Fisk and D.E. Boyce. Alternative variational inequality formulations of the network equilibrium-travel choice problem. *Transportation Science*, 17(4):454–463, Nov 1983.

[95] C.S. Fisk. Some developments in equilibrium traffic assignment. *Transportation Research*, 14B:243–255, 1980.

[96] R. Fletcher. *An Efficient, Globally Convergent, Algorithm for Unconstrained and Linearly Constrained Optimization Problems.* TP 431, AERE, Harwell, 1971.

[97] M. Florian and M. Gaudry. A conceptual framework for the supply side in transportation systems. *Transportation Research*, 14B:1–8, 1980.

[98] J. Fortuny-Amat and B. McCarl. A representation and economic interpretation of a two level programming problem. *Journal of Operation Research Society*, 32(9):783–792, 1981.

[99] M. Frank and P. Wolfe. An algorithm for quadratic programming. *Naval Research Logistics Quarterly*, 3:95–110, 1956.

[100] T.L. Friesz. Transportation network equilibrium, design and aggregation: Key developments and research opportunities. *Transportation Research*, 19A:413–427, 1985.

[101] T.L. Friesz, J. Gottfried, and E.K. Morlok. A freight network equilibrium model. 1981. Paper presented at the Transportation Equilibrium Supply Models Symposium.

[102] T.L. Friesz and P.T. Harker. Properties of the iterative optimization-equilibrium algorithm. *Civil Engineering System*, 2:142–154, 1985.

[103] T.L. Friesz and R.L. Tobin. Formaulating and solving the derived demand network equilibrium in terms of arc variables. *Journal of Regional Science*, 23:187–198, 1983.

[104] T.L. Friesz, R.L. Tobin, and P.T. Harker. Predictive intercity freight network models: the state of the art. *Transportation Research*, 17A:409–417, 1983.

[105] William L. Garrison. Basic research on transportation facilities. *Transportation Research*, 19A:511–518, 1985.

[106] N.H. Gartner. Optimal traffic assignment with elastic demands: A review; part I, analysis framework; part II, algorithmic approaches. *Transportation Science*, 14(2):174–208, 1980.

[107] A. Geoffrion. Elements of large scale mathematical programming. *Management Science*, 16, July 1970.

[108] J.B. Goddard. Technology forecasting in spatial context. *Futures*, 12:90–105, 1980.

[109] B.L. Golden and E.K. Baker. Future directions in logistics research. *Transportation Research*, 19A:405–409, 1985.

[110] A. Goodall. *The Guide to Expert Systems.* Learned Information Ltd., Oxford, England, 1985.

[111] J. Green. *Western energy: The interregional coal analysis model.* REPORT Technical bulletin 1627, U.S. Department of Agriculture, Washington, D.C., 1980.

[112] A.E. Haghani and M.S. Daskin. A short-term rail freight optimization model. 1984. Paper presented at the Dallas ORSA/TIMS Meeting.

[113] S.Y. Han and T.J. Kim. Can expert systems help with planning? *Journal of American Planning Association*, 296–308, Summer 1989.

[114] P.T. Harker. *Prediction Intercity Freight Flows: Theory and Application of a Generalized Spatial Price Equilibrium Model.* PhD thesis, University of Pennsylvania, Philadelpia, PA, 1983.

[115] P.T. Harker and T.L. Friesz. Bounding the solution of the continuous equilibrium network design problem. *Proceedings of the 9th International Symposium on Transporation and Traffic Theory*, 233–252, 1984.

[116] C. Harris. *Regional Economic Effects of Alternative Highway Systems.* Ballinger, Cambridge, MA, 1974.

[117] A.C. Hax and H.C. Meal. Hierarchical integration of production planning and scheduling. In M.A. Geisler, editor, *Logistics*, North-Holland, 1975.

[118] G.J.D. Hewings. Regional and interregional interdependencies: Alternative accounting systems. *Environment and Planning A*, 14:1587–1600, 1982.

[119] G.J.D. Hewings. Regional planing: Problems in the application of interregional input-output analysis to state planning and program activities. *Annals of Regional Science*, 4:114–122, 1970.

[120] Highway Research Board. *Highway Capacity Manual.* National Academy of Sciences, National Research Council, Washington, D.C., special report 87 edition, 1965.

[121] D.M. Himmelblau. *Applied Nonlinear Programming.* McGraw-Hill, New York, 1972.

[122] R. Hooke and T.A. Jeeves. Direct search solution of numerical and statistical problems. *J. Assoc. Computer Mach.*, 8(212), 1962.

[123] J.P. Ignizio. *Goal Programming and Extensions.* D.C. Heath and Company, Lexington, MA, 1976.

[124] N.A. Irwin, N. Dodd, and H.G. Von Cube. Capacity restraint in assignment programs. *Highway Research Board Bulletin*, 297:109–127, 1961.

[125] N.A. Irwin and H.G. Von Cube. Capacity restraint in multi-travel mode assignment programs. *Highway Research Board Bulletin*, 347:258–289, 1962.

[126] W. Isard. Interregional and regional input-output analysis: A model of a space economy. *Review of Economics and Statistics*, 33:318–328, 1951.

[127] H. Isermann. Proper efficiency and the linear vector maximum problem. *Operation Research*, 22:189–191, 1974.

[128] R.G. Jeroslow. The polynomial hierarchy and a simple model for competitive analysis. *Mathematical Programming*, 32:146–164, 1985.

[129] W.C. Jordan and L.D. Burns. Truck back-hauling on two terminal networks. *Transportation Research*, 18B:487–503, 1984.

[130] W.C. Jordan and M.A. Turnquist. A stochastic dynamic model for railroad car distribution. *Transportation Science*, 17:123–145, 1983.

[131] R. L. Keeney and H. Raiffa. *Decisions with Multiple Objectives : Preferences and Value Trade-Offs.* Wiley, New York, 1976.

[132] E.W. Kim. *Historical Research on Korean Physical Developments (Hankuk Gookto Gaebalsa Yeongoo).* Daehak-Doseo, Seoul, Korea, 1982.

[133] J.G. Kim. Population concentrationa and strategy for decentralization. 1980. Paper presented at the Second World Congress of Arts and Sciences, June 4-15.

[134] T.J. Kim. Alternative transportation modes in a land use model: a general equilibrium approach. *Journal Urban Economics*, 6:197–215, 1979.

[135] T.J. Kim. A combined land use-transportation model when zonal travel demand is endogenously determined. *Transportation Research*, 17B:449–462, 1983.

[136] T.J. Kim. *Integrated Urban Systems Modeling: Theory and Applications.* Kluwer Academic Publishers, Norwell, MA, 1989.

[137] T.J. Kim. Modeling the density variations of urban land uses with transportation network congestion. *Journal of Urban Economics*, 19:264–276, 1986.

[138] T.J. Kim, D.E. Boyce, and G.J.D. Hewings. Combined input-output and commodity flow models for interregional development planning: Insights from a Korean experience. *Geographical Analysis*, 15:330–342, 1983.

[139] T.J. Kim et al. *Optimum Transport Sector Investment Study.* Final Report, Reginal Development Research Institute / Korea Institute of Science and Technology (RDRI/KIST), July 1980. Prepared for the Ministry of Transportation, Korea and World Bank.

[140] T.J. Kim and J.G. Kim. Issues in building a national transportation development model: Experience from a Korean application. *Annals of Regional Science*, 19:18–36, 1985.

[141] T.J. Kim, C.H. Park, and J.H. Rho. Modelling investment priorities for national road improvements: a case study of Korea. *Papers of the Regional Science Association*, 57:91–105, 1985.

[142] T.J. Kim and S. Suh. Toward developing a national transportation planning model: A bilevel programming approach for Korea. *The Annals of Regional Science*, XXSPED:65–80, 1988.

[143] T.J. Kim, L.L. Wiggins, and J.R. Wright. *Expert Systems: Applications to Urban Planning.* Springer-Verlag, New York, NY, 1989.

[144] S. Ko and G.J.D. Hewings. A regional computable general equilibrium model for Korea. *Korean Journal of Regional Science*, 2(1), 1986.

[145] C.D. Kolstad. Empirical properties of economic incentives and command-and-control regulations for air pollution control. *Land Economics*, 62(3):250–263, 1986.

[146] C.D. Kolstad. *A Review of the Literature on Bi-Level Mathematical Programming.* Report LA-10284-MS, Los Alamos National Laboratory, Los Alamos, NM, 1985.

[147] C.D. Kolstad and L.S. Lasdon. Derivative evaluation and computational experience with large bi-level mathematical programs. 1986.

[148] J. Kornai. Multi-level planning system. In P. Geroux, editor, *Multilevel Planning Case Studies in Mexico*, North-Holland, Amsterdam, 1973.

[149] A. L. Kornhauser et al. The princeton railroad network model: application of computer graphics in the analysis of a changing industry. 1979. Presented at the 1979 Harvard Graphics Conference.

[150] A.L. Kornhauser. *A preliminary view of U.S. coal transportation using the Princeton railroad network information system.* REPORT ANL/EES-TM-190, Argonne National Laboratory, Argonne, IL, 1982. Proceedings of Coal Transportation Modeling Workshop.

[151] D.T. Kresge and P.O. Roberts. *Systems analysis and simulation models.* The Brookings Institute, Washington, D.C., 1971.

[152] KRIHS. *National Highway Study.* Report, Korea Research Institute for Human Settlement, Seoul, Korea, 1985.

[153] A. Kumar. Impact of technological developments on urban form and travel behavior. 1989. Working Paper.

[154] R. Langendorf. Computer and decision making. *Journal of American Planning Association*, 422–433, 1985. Autumn.

[155] Z.F. Lansdowne. *Rail Freight Traffic Assignment.* Research Report DOT-TSC-OST-79, Control Analysis Corporation, Palo Alto, CA, 1979.

[156] Z.F. Lansdowne. Rail freight traffic assignment. *Transportation Research*, 15A:83–109, 1981.

[157] L.S. Lasdon. *Optimization Theory for Large Systems.* The Macmillian Company, New York, 1970.

[158] L. J. LeBlanc. An algorithm for the discrete network design problem. *Transportation Science*, 9:183–199, 1975.

[159] L.J. LeBlanc and D.E. Boyce. A bilevel programming algorithm for exact solution of the network design problem with user-optimal flows. *Transportation Research*, 20B(3):259–265, 1986.

[160] L.J. LeBlanc, E.K. Morlok, and W.P. Pierskalla. An efficient approach to solving the road network equilibrium traffic assignment problem. *Transportation Research*, 9:309–318, 1975.

[161] E.E. Lee. *A Study on Application of Link Congestion Function and Generalized Cost Function.* Master's thesis, Seoul National University, Seoul, Korea, 1987.

[162] Y.J. Lee. *Methods for Implementing Network Equilibrium Models of Urban Location and Travel Choice.* PhD thesis, University of Illinois, Urbana, IL, 1986.

[163] J.K. Lenstra, D.S. Johnson, and A.H.G. Rinnooy Kan. The complexity of the network design problem. 1976. Mathematisch Centrum, Amsterdam.

[164] W.W. Leontief and A. Strout. Multiregional input-output analysis. In T. Barna, editor, *Structural Interdependence and Economic Development*, pages 119–150, Macmillan, New York, 1963.

[165] T. Leventhal, G. Nemhauser, and L. Trotter Jr. A column generating algorithm for optimal traffic assignment. *Transportation Science*, 7:168–176, 1973.

[166] Harold A. Linstone, editor. *Multiple Perspectives for Decision-making: Bridging the Gap between Analysis and Action.* North Holland, New York, 1984.

[167] I.S. Lowry. *A Model of Metropolis.* Research Report RM-4125-RC, Rand Corporation, Santa Monica, CA, 1964.

[168] S.M. Macgill. The Lowry model as an input-output model and its extension to incorporate full intersectional relations. *Regional Studies*, 11:337–354, 1977.

[169] T.L. Magnanti and R.T. Wong. Network design and transportation planning: models and algorithms. *Transportation Science*, 18(1):1–55, February 1984.

[170] M.L. Manheim. Understanding supply in transportation systems. *Transportation Research*, 14A:119–136, 1980.

[171] D.F. Marble and W.L. Garrison. Analysis of transportation networks: A linear programming fromulation. *Highway Research Board*, 37:1–17, 1958.

[172] P. Marcotte. Network design problem with congestion effects: A case of bilevel programming. *Mathematical Programming*, 34:142–162, 1986.

[173] P. Marcotte. Network optimization with continuous control parameters. *Transportation Science*, 17:181–197, 1983.

[174] P. Marcotte. A note on a bilevel programming algorithm by LeBlanc and Boyce. *Transportation Research*, 22B(3):233–237, 1988.

[175] P. Marcotte and G. Marquis. Efficient implementation of heuristic for the continuous network design problem. 1988.

[176] P.N. Mathur. Multiregional analysis in a dynamic input-output framework. In A.P. Carter and A. Brody, editors, *Input-Output Techniques*, pages 208–230, North Holland, Amsterdam, 1972.

[177] Micro Database Systems, Inc. *GURU Reference Manual.* Micro Database Systems Inc., Lafayette, IN 47902, 1987.

[178] E. S. Mills. Markets and efficient resource allocation in urban areas. *Swedish Journal of Urban Economics*, 74:100–113, 1972.

[179] E.S. Mills. Planning and market processes in urban models. In R.E. Grieson, editor, *Public and Urban Economics: Essays in Honor of William Vickrey*, D.C. Heath & Co., Lexington, MA, 1975.

[180] Ministry of Construction. *Feasibility Study Report for ADB Fourth Road Improvement Project.* Final Report, Ministry of Construction, Republic of Korea, July 1987.

[181] Ministry of Transportation. Statistical yearbook of transportation. published annually by the Ministry of Transportation, Korea, 1986.

[182] F. Moavenzadeh. Research needs in transportation facilities: Guideway technology and materials research. *Transportation Research*, 19A:497–509, 1985.

[183] J.T. Moore and J.T. Bard. An algorithm for the zero-one bilevel programming problem. 1987.

[184] E.K. Morlok. Supply function for public transport: initial concepts and models. In M. Florian, editor, *Traffic equilibrium methods*, pages 322–367, Springer-Verlag, Berlin, 1976.

[185] E.K. Morlok, J.L. Schofer, W.P. Pierskalla, R.E. Marsten, S.K. Agarwal, J.L. Edwards, L.J. LeBlanc, D.T. Spacek, and J.W. Stoner. Development and application of a highway network design model. 1973. Prepared for Federal Highway Administration, Environment Planning Branch.

[186] L. Moses. The stability of interregional trading patterns and input-output analysis. *American Economics Review*, 45:803–832, 1955.

[187] W.W. Mosher. A capacity restraint algorithm for assigning flow to a transport network. *Highway Research Record*, 6:41–70, 1963.

[188] H. Moulin. *Game Theory for the Social Sciences.* New York University Press, New York, 1982.

[189] J. D. Murchland. Braess paradox of traffic flow. *Transportation Research*, 4:391–394, 1970.

[190] S.C. Narula and A.D. Nwosu. An algorithm to solve a two-level resource control pre-emptive hierarchical programming problem. In P. Serafini, editor, *Mathematics of Multi-Objective Optimization*, pages 253–373, Springer-Verlag, 1985.

[191] S.C. Narula and A.D. Nwosu. *A Dynamic Programming Solution for the Hierarchical Linear Programming Problem.* Research Report 37-82-10, Dept. of Operation research and Statistics, Rensselaer Polytechnic Institute, Troy, NY, 1982.

[192] S.C. Narula and A.D. Nwosu. Two-level hierarchical programming problem. In P. Hansen, editor, *Essays and Surveys on Multiple Criteria Decision Making*, pages 290 – 299, Springler-Verlag, Berlin, 1983.

[193] J. Nash. Non-cooperative games. *Annals of Mathematics*, 54:286–295, 1951.

[194] E.T. Nevin, A.R. Roe, and J.I. Round. *The structure of the Welsh economy.* Cardiff, Welsh, 1966.

[195] S. Nguyen. *Une Approche Unifiee des Methods d'equilibre pour l'affectation du Traffic.* PhD thesis, Department d'Informatique, Universite de Montreal, Quebec, 1974.

[196] L.Y. Nicol. Communications technology: Economic and spatial impacts. In M. Castells, editor, *High Technology, Space, and Society*, Sage Publications, Inc., 1985.

[197] A.D. Nwosu. *Pre-emptive Hierarchical Programming Problem ; A Decentralized Decision Model.* PhD thesis, Rensselaer Polytechnic Institute, Troy, NY., 1983.

[198] I. Okutani. Equilibrium flows in a network with congested links. *Proceedings of the 9th International Symposium on Transportation and Traffic Theory*, 253–271, 1984.

[199] K.R. Overgaard. Urban transportation planning: Traffic estimation. *Traffic Quarterly*, 197–218, 1967.

[200] E.R. Peterson and H.V. Fullerton. *The Railcar Network Models.* Technical Report, Canadian Institute of Guided Transport, Kingston, Canada, 1975.

[201] K.R. Polenske. Empirical implementation of a multiregional input-output gravity trade model. In A.P. Carter and A. Brody, editors, *Contributions to Input-Output Analysis*, pages 143–163, North Holland, Amsterdam, 1970.

[202] K.R. Polenske. An empirical test of interregional input-output model: Estimation of 1963 Japanese production. *American Economic Review*, 60:76–82, 1970.

[203] K.R. Polenske. The implementation of a multiregional input-output model for the U.S. In A.P. Carter and A. Brody, editors, *Input-Output Techniques*, pages 143–163, North Holland, Amsterdam, 1972.

[204] M.J.D. Powell. Extensions to subroutine VF02AD. In R.F. Drenick and F. Kozin, editors, *System Modeling and Optimization*, pages 529–539, Springer-Verlag, Berlin, 1982. Lecture Notes in Control and Information Science, vol. 38.

[205] W.B. Powell, Y. Sheffi, and S. Thiriez. The dynamic vehicle allocation problem with uncertain demands. In *Ninth International Symposium on Transportation and Traffic Theory*, pages 357–374, VNU Science Press, Utrecht, The Netherlands, 1984.

[206] G. Pyatt and A.R. Roe. *Social Accounting for Development Planning.* Cambridge University Press, New York, NY, 1977.

[207] R.F. Quandt. Models of transportation and optimal network construction. *Journal of Regional Science*, 2:27–46, 1962.

[208] B.A. Ralston and G.M. Barber. Taxation and optimal road penetration. *Geographical Analysis*, 16:313–330, 1984.

[209] B.A. Ralston and G.M. Barber. A theoretical model of road development dynamics. *Annals of the Association of American Geographers*, 72:201–210, 1982.

[210] H. Richardson. Monocentric vs. policentric models: The future of urban economics in regional science. 1988. Presidential address to the Western Regional Science Association Twenty-Seventh Annual Meeting, Napa Valley, CA.

[211] T.M. Ridley. An investment policy to reduce the travel time in a transportation network. *Transportation Research*, 2:409–424, 1968.

[212] P. Rietveld. A review of multiregional economic models. November 1981. Paper presented at the Conference on Practice and Prospects of Multiregional Economic Modeling at the International Institute for Applied Systems Analysis.

[213] P.O. Roberts. *Transport Planning: Models for Developing Countries.* PhD thesis, Northwestern University, Evanston, IL, 1966.

[214] V.B. Robinson, A.U. Frank, and M.A. Blaze. Expert systems applied to problems in geographic information systems: introduction, review, and prospects. *Computer, Environment, and Urban Systems*, 11(9):161–173, 1986.

[215] E.M. Rogers. *Communications Technology: The new media in society.* The Free Press, Macmillan, Inc., 1986.

[216] E. Romanoff. The economic base model: A very special case of input-output analysis. *Journal of Regional Science*, 14:121–129, 1974.

[217] J. I. Round. Regional input-output models in the United Kindom: A reappraisal of some techniques. *Regional Studies*, 6:1–9, 1972.

[218] SCAI. *PCGPAK User's Guide.* Scientific Computing Association, Inc., New Haven, CT, version 1.04 edition, October 1987.

[219] G. Schenk. *A Multilevel Programming Model for Determining Regional Effluent Charges.* Master's thesis, Department of Industrial Engineering, SUNY at Buffalo, 1980.

[220] J.L. Schofer and D.E. Boyce. Conference summary and conclusions. *Transportation Research*, 19A:351–354, 1985.

[221] A.J. Scott. The optimal network problem: Some computational procedures. *Transportation Research*, 3:201–210, 1969.

[222] F. Seo and M. Sakawa. *Evaluation for Industrial Land- Use Program Related to Water Quality Management.* Working Paper 80- 49, International Institute for Applied Systems Analysis, Luxemburg, Austria, 1980.

[223] J.P. Shaw. *A Parametric Complementary Pivot Approach to Multilevel Programming.* Master's thesis, Dept. of Industrial Engineering, SUNY at Buffalo, 1980.

[224] Y. Sheffi. *Urban Transportation Networks: Equilibrium Analysis with Mathematical Programming Methods.* Prentice-Hall, Englewood Cliffs, NJ, 1985.

[225] Y. Sheffi and C.F. Daganzo. Computation of equilibrium over transportation networks: The case of disaggregate demand models. *Transportation Science*, 14(2):155–173, 1980.

[226] K. Shimizu and E. Aiyoshi. A new computational method for stackelberg and min-max problems by use of a penalty method. *IEEE Transactions on Automatic Control*, AC-26(2):460–466, 1981.

[227] M. Simaan and J.B. Cruz Jr. On the stackelberg strategy in nonzero-sum games. *Journal of Optimization Theory and Applications*, 11(5):533–555, 1973.

[228] M. J. Smith. The existence, uniqueness and stability of traffic equilibria. *Transportation Research*, 13B:293–304, 1979.

[229] R.J. Smock. An iterative assignment approach to capacity restraint on arterial networks. *Highway Research Board Bulletin*, 347:60–66, 1962.

[230] T.J. Soltman. Effects of alternate loading sequences on results from Chicago trip distribution and assignment model. *Highway Research Record*, 114:122–140, 1965.

[231] S. Stair. Selecting an optimal traffic network. *Journal of Transport Economics and Policy*, 2:218–231, 1967.

[232] T.M. Stanback. The changing fortunes of metropolitan economics. In M. Castells, editor, *High Technology, Space, and Society*, Sage Publications, Inc., 1985.

[233] P.A. Steenbrink. *Optimization of Transport Network.* John Wiley and Sons, New York, NY, 1974.

[234] R. Stone et al. *A programme for growth.* Cambridge, 1963.

[235] R.E. Strauch. *A Critical Assessment of Quantitative Methodology as a Political Analysis Tool.* Land Corporation, Santa Monica, 1974.

[236] S. Suh. *Implementation and Evaluation of Nonlinear Bilevel Programming Model of Equilibrium Network Design Problem.* PhD thesis, University of Illinois at Urbana-Champaign, Urbana, IL, January 1989.

[237] S. Suh and T.J. Kim. A solution for nonlinear bilevel programming models of the equilibrium network design problem. 1989. Department of Urban and Regional Planning, University of Illinois, Urbana, IL.

[238] S. Suh and T.J. Kim. Solving a nonlinear bilevel programming model of equilibrium network design problem for Korea. 1988. Forthcoming in the Papers of the Regional Science Association: Stockholm Volume.

[239] S. Suh, M.P. Park, and T.J. Kim. ESMAN: an expert system for manufacturing site selection. *Computers, Environment, and Urban Systems*, 12:239–252, 1989.

[240] S.M. Suh. Revamping the five-year plan: Issues and dilemmas. *Business Korea*, 1-3:62–63, 1983.

[241] C. Suwansirikul, T.L. Friesz, and R.L. Tobin. Equilibrium decomposed optimization: A heuristic for the continuous equilibrium network design problem. *Transportation Science*, 21(4):254–263, 1987.

[242] C. Swerdloff. Developing a national network model of intercity freight movement in the United States. 1971. Proceedings of the International Symposium Freight Traffic Models.

[243] H-N. Tan, S.B. Gershwin, and M. Athans. *Hybrid Optimization in Urban Traffic Networks.* Report DOT-TSC-RSPA-79-7, Massachusetts Institute of Technology, Laboratory for Information and Decision Systems, Cambridge, MA, 1979.

[244] M. A. P. Taylor. Parameter estimation and sensitivity of parameter values in a flow-rate/travel-time relation. *Transportation Science*, 11(3):275–292, 1977.

[245] R. L. Tobin and T. L. Friesz. Sensitivity analysis for equilibrium network flow. *Transportation Science*, 22(4):242–250, Nov. 1988.

[246] R.L. Tobin. Sensitivity analysis for variational inequalities. *Journal of Optimization Theory and Application*, 48(1):191–204, 1986.

[247] Traffic Research Corporation. *Winnipeg Area Transportation Study.* Technical Report, 1966. Report Prepared for the Streets and Transit Division of the Metropolitan Corporation of Greater Winnipeg.

[248] Transportation Research Board. *Highway Capacity Manual.* National Research Council, Washington, D.C., special report 209 edition, 1985.

[249] M.A. Turnquist. Research opportunities in transportation system characteristics and operations. *Transportation Research*, 19A:357–366, 1985.

[250] M. Wachs. Planning, organizations and decision-making: A research agenda. *Transportation Research*, 19A(5/6):521–531, 1985.

[251] J.G. Wardrop. Journey speed and flow in central London. *Traffic Engineering and Control*, 9:528–532, 1968.

[252] J.G. Wardrop. Some theoretical aspects of road traffic research. *Proceedings of Institution of Civil Engineering*, Part II, 1:325–378, 1952.

[253] D. Waterman. *A Guide to Expert Systems.* Addison-Wesley Publishing Company, Massachusetts, 1986.

[254] M.P. Wellman. Reasoning about assumptions underlying mathematical models. In J.S. Kowalik, editor, *Coupling Symbolic and Numerical Computing in Expert Systems*, Elsevier Science Publishers B.V., New York, 1986.

[255] U.P. Wen. The 'kth-best' algorithm for multilevel programming. 1980.

[256] U.P. Wen. *Mathematical Methods for Multilevel Linear Programming.* PhD thesis, State University of New York at Buffalo, NY, 1981.

[257] C. Werner. The role of technology and geometry in optimal network design. *Papers of Regional Science Association*, 21:173–190, 1968.

[258] I.N. Williams. Algorithm 1: three point rational function interpolation for calibrating gravity models. *Environment and Planning A*, 9:215–221, 1977.

[259] A. G. Wilson. *Entropy in Urban and Regional Modelling.* Pion Ltd., London, 1970.

[260] A. G. Wilson. Interregional commodity flows: Entropy maximizing approaches. *Geographical Analysis*, 2:255–282, 1970.

[261] A. G. Wilson, J.C. Coelho, S.M. Macgill, and H.C.W.L. Williams. *Optimization in Locational and Transport Analysis.* Wiley, New York, NY, 1981.

[262] R.T. Wong. Introduction and recent advances in network design. In M. Florian, editor, *Transportation Planning Models*, North Holland, Amsterdam, 1983.

[263] C.F. Wood. Application of direct search to the solution of engineering problem. 1960.

[264] D.M. Wood and J.R. Wright. Calibrating complex simulation models using rules-based inferencing. 1987.

[265] W.I. Zangwill and C.B. Garcia. *Pathways to Solution, Fixed Points and Equilibria.* Prentice-Hall, Englewood Cliffs, NJ, 1981.

[266] Z. Zlatev, J. Wasniewski, and K. Schaumberg. Y12m: solution of large and sparse systems of linear algebraic equations. 1980.

Author Index

A

B

C

S

T

V

W

Z

Subject Index

M

N